SpringerBriefs in Physics

For further volumes:
http://www.springer.com/series/8902

Jie Liu

Classical Trajectory Perspective of Atomic Ionization in Strong Laser Fields

Semiclassical Modeling

 Springer

Jie Liu
Institute of Applied Physics
and Computational Mathematics
Beijing
People's Republic of China

ISSN 2191-5423   ISSN 2191-5431   (electronic)
ISBN 978-3-642-40548-8   ISBN 978-3-642-40549-5   (eBook)
DOI 10.1007/978-3-642-40549-5
Springer Heidelberg New York Dordrecht London

Library of Congress Control Number: 2013947363

Printed on acid-free paper

Springer is part of Springer Science+Business Media (www.springer.com)

Preface

Light–matter interaction is a topic with a long history which constantly attract much attentions in modern physics. The discovery of the photoelectric effect is a milestone, in which electrons are emitted from solids, liquids, or gases when they are irradiated by light. The story can be traced back to 1887, when Heinrich Hertz discovered that electrodes illuminated with ultraviolet light create electric sparks more easily. In 1905, Albert Einstein successfully explained experimental data from the photoelectric effect as a result of light energy being carried in discrete quantized packets. Study of the photoelectric effect led to important steps in understanding the quantum nature of light and electrons and influenced the formation of the concept of wave-particle duality. This discovery led to the quantum revolution and Einstein was awarded the Nobel Prize in 1921 for "his discovery of the law of the photoelectric effect".

In Hertz's experiment, it was found that, if the photon energy is too low, the electron is unable to escape the material. The energy of the emitted electrons does not depend on the intensity of the incoming light, but only on the energy or frequency of the individual photons. However, until the end of the last century when optical technique progress provided people with a new sort of coherent and brilliant light source, i.e., intense laser beam, something changes. When noble gases were irradiated by an intense laser beam as in the experiment performed by P. Agostini, et al. in 1979, it was found that, in contrast to Hertz's, an atom can absorb multiple photons simultaneously and even more than the required number of photons for ionization. The above striking phenomenon was termed as above threshold ionization (ATI). It breaks the long-standing ionization picture of traditional perturbative theory and indicates the coming of strong-field times with the characteristic of non-perturbative phenomena .

Atomic ionization plays a fundamental role in light–matter interaction. Since the exquisite experiment in 1979, great progress has been made about atomic ionization issue from both experimental and theoretical sides. The atoms and molecules in strong laser fields have demonstrated many intriguing and complex behaviors and become an active field in modern physics. There are versatile applications in attosecond physics, X-ray generation, inertial confined fusion (ICF), and so on. In this book I will present some basic concepts and discuss some interesting topics using a semiclassical model of classical trajectory ensemble simulation. Our discussions focus on long wavelength limit for laser and tunneling ionization become an dominating mechanism. In contrast to quantum treatments, we notice that the classical trajectory approaches can revive in this situation because continuum–continuum transition plays a crucial role in strong-field ionization. The classical trajectory model has advantages of clear picture, feasible computing, and can even account for correlated electron observations quantitatively. I will introduce semiclassical tunneling ionization and present some applications of the model in such as single ionization, double ionization, neutral atom acceleration, and other timely issues in strong field physics, which can deliver useful messages to readers with providing simple classical trajectory perspective on complex atomic ionization process.

I am indebted to my family's constant support when I am writing this book. I would like to thank my students Qinzhi Xia, Kaiyun Huang, Xinfang Song, and Di-Fa Ye for help me collecting materials, checking formulas, and other paper work.

Beijing, March 2013   Jie Liu

Contents

Chapter 1
Tunneling Ionization and Classical Trajectory Model

Abstract In this chapter, we introduce the basic concepts of atomic tunneling ionization and formulate atomic ionization rate and transversal velocity distribution within the tunneling framework. We then present how to simulate the atomic ionization dynamics in intense laser fields using classical trajectory ensemble with tunneling allowance simultaneously.

1.1 Tunneling Ionization Theory

In the 1940s, L. D. Landau and E. M. Lifshitz [1] analyzed the tunneling ionization of Hydrogen atom in static electric field in parabolic coordinates. Then in the 1960s, A.M. Perelomov,V.S. Popov, and V.M. Teren'ev [2] used the "imaginary time" method to derive the formulae of atomic ionization rate and transversal velocity distribution. After two decades, M. V. Ammosov, N. B. Delone, and V. P. Krainov [3] obtained more general and quantitative results called as ADK formula. Besides, many other people contributed in this field [4]. In this section, we will discuss the atomic tunneling ionization following their spirit.

1.1.1 Hydrogen Atoms in Static Electric Field: Parabolic Coordinate

The separation of the variables in Schrödinger's equation written in parabolic coordinates is useful, especially in problems where a certain direction in space is distinctive. A typical example is the atom in an external field. The parabolic coordinates ξ, η, ϕ are defined as follows:

$$x = \sqrt{\xi\eta}\cos\phi \quad y = \sqrt{\xi\eta}\sin\phi \quad z = \frac{1}{2}(\xi - \eta) \tag{1.1}$$

J. Liu, *Classical Trajectory Perspective of Atomic Ionization in Strong Laser Fields*, SpringerBriefs in Physics, DOI: 10.1007/978-3-642-40549-5_1, © The Author(s) 2014

Fig. 1.1 The parabolic coordinates. The solid curves and
the dashed curves represent the
hyperbolas of $\xi = constant$
and $\eta = constant$, respectively

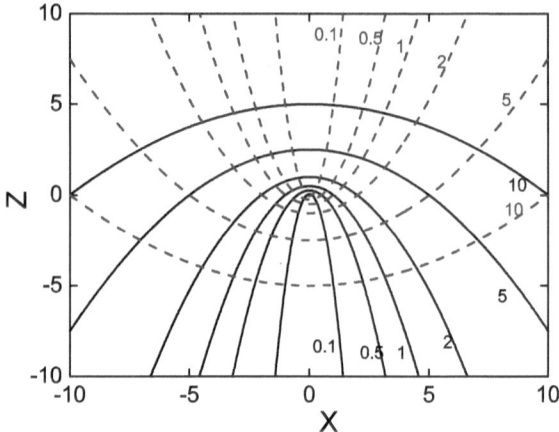

or conversely $\xi = r + z, \eta = r - z, \phi = tan^{-1}(y/x)$, where ξ and η take values
from 0 to ∞, and ϕ from 0 to 2π. The surfaces $\xi = constant$ and $\eta = constant$
are paraboloids of revolution about the z-axis, with focus at the origin (see Fig. 1.1).

Schrödinger's equation for the hydrogen atom in a uniform electric field is of
the form (Hereafter, atomic units, i.e., $m_e = e = \hbar = 1.a.u.$ are used without
specifying),

$$\left(\frac{1}{2}\nabla^2 - I_p + \frac{1}{r} - Ez\right)\psi = 0,$$

where, $m_e, e, \hbar$ are electron mass, electron charge, and Planck constant, respectively;
I_p is atomic ground state energy or ionization potential; E is instantaneous electric
field.

Let us seek the eigenfunctions ψ in the form [1]

$$\psi = f_1(\xi)f_2(\eta)e^{im\phi} \tag{1.2}$$

Then we can get the two equations:

$$\frac{d}{d\xi}\left(\xi\frac{df_1}{d\xi}\right) + \left(-\frac{1}{2}I_p\xi - \frac{1}{4}\frac{m^2}{\xi} - \frac{1}{4}E\xi^2\right)f_1 = -\beta_1 f_1 \tag{1.3}$$

$$\frac{d}{d\eta}\left(\eta\frac{df_2}{d\eta}\right) + \left(-\frac{1}{2}I_p\eta - \frac{1}{4}\frac{m^2}{\eta} + \frac{1}{4}E\eta^2\right)f_2 = -\beta_2 f_2 \tag{1.4}$$

where, $\beta_1 + \beta_2 = 1$. By substitution

$$f_1 = \frac{\chi_1}{\sqrt{\xi}}, \quad f_2 = \frac{\chi_2}{\sqrt{\eta}} \tag{1.5}$$

we transfer Eq. (1.3),(1.4) to the form

$$\frac{d^2\chi_1}{d\xi^2} + \left(-\frac{1}{2}I_p + \frac{\beta_1}{\xi} - \frac{m^2-1}{4\xi^2} - \frac{1}{4}E\xi\right)\chi_1 = 0, \tag{1.6}$$

$$\frac{d^2\chi_2}{d\eta^2} + \left(-\frac{1}{2}I_p + \frac{\beta_2}{\eta} - \frac{m^2-1}{4\eta^2} + \frac{1}{4}E\eta\right)\chi_2 = 0 \tag{1.7}$$

Each of these equations takes the form of one-dimensional Schrödinger's equation, where the energy of the particle is taken as $\frac{E}{4}$, and the potential energy of the functions is taken as

$$U_1(\xi) = -\frac{\beta_1}{2\xi} + \frac{m^2-1}{8\xi^2} + \frac{1}{8}E\xi \tag{1.8}$$

$$U_2(\eta) = -\frac{\beta_2}{2\eta} + \frac{m^2-1}{8\eta^2} - \frac{1}{8}E\eta \tag{1.9}$$

It can be seen that there is a potential barrier "along the η coordinate." For the ground state of hydrogen atom, we approximate that

$$I_p = \frac{1}{2}, \quad \beta_1 = \beta_2 = \frac{1}{2}. \tag{1.10}$$

We can suppose that the tunneling electron releases at the boundary of the barrier of $\eta = \eta_t$, satisfying $-\frac{I_p}{4} = U_2(\eta_t)$.

As the probability of ionization of hydrogen atom is concerned, we start from the wave function of the ground state in the absence of the field , i.e.,

$$\psi = \frac{1}{\sqrt{\pi}}e^{-(\xi+\eta)/2}. \tag{1.11}$$

When the field is present, the dependence of ψ on ξ in the region where we are interested can be regarded as being the same as in Eq. (1.11), while to determine its dependence on η we have Eq. (1.7), i.e.,

$$\frac{d^2\chi_2}{d\eta^2} + \left(-\frac{1}{4} + \frac{1}{2\eta} + \frac{1}{4\eta^2} + \frac{1}{4}E\eta\right)\chi_2 = 0 \tag{1.12}$$

so that

$$\psi = \frac{\chi_2}{\sqrt{\eta}}e^{-\frac{1}{2}\xi} \tag{1.13}$$

Let η_0 be some value of η (within the barrier) such that $1 \ll \eta_0 \ll \frac{1}{E}$. For $\eta > \eta_0$, the wave function is quasi-classical. So that we can use the results of the quasi-classical case and the boundary condition that ψ must become the wave function Eq. (1.11) at $\eta = \eta_0$. We obtain in the region outside the barrier the expression

$$\chi_2(\eta) = \left(\frac{\eta_0 |p(\eta_0)|}{\pi p(\eta)}\right)^{1/2} \exp\left(-\frac{\eta_0}{2} + i\int_{\eta_0}^{\eta} p d\eta + \frac{1}{4}i\pi\right), \tag{1.14}$$

where

$$p(\eta) = \sqrt{-\frac{1}{4} + \frac{1}{2\eta} + \frac{1}{4\eta^2} + \frac{E\eta}{4}} \tag{1.15}$$

We shall be interested only in the square of χ_2. Hence the imaginary part of the exponent is unimportant. Denoting by η_1 the root of the equation $p(\eta) = 0$, we have

$$\begin{aligned}
|\chi_2|^2 &= \frac{\eta_0 |p(\eta_0)|}{\pi p(\eta)} \exp\left(-2\int_{\eta_0}^{\eta_1} |p(\eta)|d\eta - \eta_0\right) \\
&= \frac{4}{\pi E} \frac{e^{-2/3E}}{\sqrt{E\eta - 1}}
\end{aligned} \tag{1.16}$$

i.e.

$$|\psi|^2 = \frac{4}{\pi E\eta\sqrt{E\eta - 1}} \exp\left(-\xi - 2/3E\right), \tag{1.17}$$

The ionization probability w can be calculated by the current through the plane perpendicular to the z-axis as follows:

$$w = \int_{-\infty}^{\infty}\int_{-\infty}^{\infty} |\psi|^2 v_z dx dy, \tag{1.18}$$

where v_z the velocity of electron at large z distance can be calculated by the energy relation:

$$\frac{1}{2}v_x^2 + \frac{1}{2}v_y^2 + \frac{1}{2}v_z^2 + Ez = -I_p. \tag{1.19}$$

Suppose $v_z \gg v_x$, $v_z \gg v_y$ and $\eta \gg \xi$, we then have,

$$v_z = -\sqrt{E\eta - 2I_p} = -\sqrt{E\eta - 1}. \tag{1.20}$$

Since

$$\rho = \sqrt{x^2 + y^2} = \sqrt{\xi\eta}, \tag{1.21}$$

keeping η as constant when $\eta \gg 0$, we have

$$d\rho = \sqrt{\eta}\frac{d\xi}{2\sqrt{\xi}}. \tag{1.22}$$

So the tunneling probability is

$$w = \int_{-\infty}^{\infty} \int_{-\infty}^{\infty} |\psi|^2 v_z dx dy = \frac{4}{E} \exp(-\frac{2}{3E}). \qquad (1.23)$$

The distribution of transverse velocity $v_\perp$ ($v_\perp^2 = v_x^2 + v_y^2$) can be calculated as the following. Considering the classical case in the constant external field along z-axis, the coordinate ξ of the electron is

$$\xi = r + z = \frac{x^2 + y^2}{\sqrt{x^2 + y^2 + z^2} - z}$$

$$= \frac{v_\perp^2 t^2}{\sqrt{v_\perp^2 t^2 + (\frac{1}{2}Et^2)^2} + \frac{1}{2}Et^2}$$

$$= \frac{v_\perp^2}{E} \quad (t \longrightarrow \infty) \qquad (1.24)$$

Since the distribution of ξ is

$$f(\xi) \sim e^{-\xi}, \qquad (1.25)$$

the distribution of transverse velocity of the electron after tunneling ionization is

$$f(v_\perp) \sim \frac{2v_\perp}{E} e^{-\frac{v_\perp^2}{E}}. \qquad (1.26)$$

1.1.2 Landau-Dyhne Adiabatic Approximation

The transition amplitude A_{if} from the initial state i to the final state f according to the Landau-Dyhne adiabatic approximation has the form [3]

$$A_{if} = \exp(\frac{i}{\hbar} \int_0^{t_0} (\varepsilon_f(t) - \varepsilon_i(t)) dt), \qquad (1.27)$$

where $\varepsilon_i(t)$ and $\varepsilon_f(t)$ are the adiabatic energy of the initial and final atomic states, respectively. The time t_0 is the complex classical turning point, where

$$\varepsilon_f(t_0) = \varepsilon_i(t_0). \qquad (1.28)$$

Let

$$S = \int_0^{t_0} (\varepsilon_f(t) - \varepsilon_i(t)) dt, \qquad (1.29)$$

then the transition rate from initial state i to final state f is

$$w_{if} = |A_{if}|^2 = \exp\left(-\frac{2}{\hbar} Im\, S\right). \tag{1.30}$$

In atomic unit, the transition rate is written as

$$w_{if} = \exp\left(-2Im\, S\right). \tag{1.31}$$

So what we need to calculate is $Im\, S$.

First, assuming the initial state i is the hydrogen ground state and slightly perturbed by the external field, we write

$$\varepsilon_i(t) = -I_p = -1/2.$$

Second, after the variables separated in the parabolic coordinates as mentioned in the above section, the effective potential is

$$U_{eff}(\eta) = -(2I_p)^{1/2}/\eta - E(t)\eta/2, \tag{1.32}$$

where we set the electric field is static along z axis,

$$\mathbf{E}(t) = \hat{e}_z E \tag{1.33}$$

Then in velocity gauge, the Hamiltonian of the system is

$$H(t) = \frac{1}{2}(\mathbf{p}(t) - e\mathbf{A}(t))^2 - (2I_p)^{1/2}/\eta \tag{1.34}$$

where $\mathbf{p}(t)$ is the generalized momentum, and

$$\mathbf{A}(t) = -\hat{e}_z E t, \tag{1.35}$$

so that the final energy

$$\varepsilon_f(t) = H(t) = \frac{1}{2}((\mathbf{p}(t) - \hat{e}_z E t)^2 + p_\perp^2) - (2I_p)^{1/2}/\eta. \tag{1.36}$$

Ignoring the Coulomb potential, from $\varepsilon_f(t_0) = \varepsilon_i(t_0)$ we obtain that

$$t_0 = (p_z + i(2I_p + p_\perp^2)^{\frac{1}{2}})/E \tag{1.37}$$

According to the equation of motion without Coulomb force in parabolic coordinates, we have

$$\dot{\eta} = \frac{4p_\eta \eta}{m(\eta + \xi)} \approx 4p_\eta \quad (\xi \sim 0) \tag{1.38}$$

$$\dot{p}_\eta = \frac{E(t)}{2} + \frac{p_\varphi^2}{2m\eta^2\xi} - \frac{2p_\eta^2}{m(\eta + \xi)} + \frac{2(p_\eta^2\eta + p_\xi^2\xi)}{m(\eta + \xi)^2} \approx \frac{E}{2}. \tag{1.39}$$

Since $\frac{\xi - \eta}{2} = z, \dot{\eta} = -2\dot{z}, t = 0$, let $\dot{z} = i(2I_p)^{1/2}$ (ignoring $p_\perp$ when calculating $\eta(t)$),

$$\dot{\eta}(0) = -i(8I_p)^{1/2} \tag{1.40}$$

$$p_\eta(0) = \dot{\eta}(0)/4 = -i(2I_p)^{1/2}/2 \tag{1.41}$$

So

$$p_\eta(t) = -i(2I_p)^{1/2}/2 + \frac{E}{2}t \tag{1.42}$$

$$\eta(t) = \eta_2 - i(8I_p)^{1/2}t + Et^2 \tag{1.43}$$

The calculation of S is partitioned into two parts [2]:

$$\begin{aligned}
S &= \int_0^{t_0} (\varepsilon_f(t) - \varepsilon_i(t)) dt \\
&= \int_0^{t_1} (\varepsilon_f(t) - \varepsilon_i(t)) dt + \int_{t_1}^{t_0} (\varepsilon_f(t) - \varepsilon_i(t)) dt \\
&= S_1 + S_0
\end{aligned} \tag{1.44}$$

Here t_1 is a small parameter. And after the summation of S_1 and S_0, we acquire the limitation of $t_1 \to 0$.

Approximating $\eta_2 \sim 0$, we calculate S_0 as follows:

$$\begin{aligned}
S_0 &\approx \int_{t_1}^{t_0} (\frac{1}{2}((p_z - Et)^2 + p_\perp^2) + I_p - \frac{(2I_p)^{1/2}}{-i(8I_p)^{1/2}t + Et^2}) dt \quad (\eta_2 \to 0) \\
&= S_{01} + S_{02}.
\end{aligned} \tag{1.45}$$

For the first term:

$$\begin{aligned}
S_{01} &= \int_{t_1}^{t_0} (\frac{1}{2}(p_z - Et)^2 + (I_p + \frac{1}{2}p_\perp^2)) dt \\
&= ((\frac{1}{2}p_z^2 + (I_p + \frac{1}{2}p_\perp^2))t - \frac{1}{2}Ep_z t^2 + \frac{E^2}{6}t^3)|_{t_1}^{t_0}.
\end{aligned} \tag{1.46}$$

Since $t_1 \to 0$, the lower limit of integral is real. As what we are concerned with is only the imaginary part, so

$$ImS_{01} = Im(\frac{1}{2}(p_z^2 + 2(I_p + \frac{1}{2}p_\perp^2))t_0 - \frac{1}{2}Ep_zt_0^2 + \frac{E^2}{6}t_0^3). \qquad (1.47)$$

Take t_0 which in Eq. 1.37 into the above formula, we get

$$Im S_{01} = \frac{(2I_p + p_\perp^2)^{\frac{3}{2}}}{3E}. \qquad (1.48)$$

This result is consistent with Ref. [3], except that the high order term here disappears because static field is exploited here.

Then we calculate $Im S_{02}$. Here we only take the first order effect into account, so we have $t_0 = i(2I_p)^{1/2}/E$ and $t_1 = \frac{\eta_1}{-i(8I_p)^{1/2}}$. So

$$\begin{aligned}
S_{02} &= \int_{t_0}^{t_1} -\frac{(2I_p)^{1/2}}{-i(8I_p)^{1/2}t + Et^2}dt \\
&= -\frac{i(2I_p)^{1/2}}{(8I_p)^{1/2}} \ln\left(\frac{t_0}{t_1}\frac{Et_1 - i(8I_p)^{1/2}}{Et_0 - i(8I_p)^{1/2}}\right) \\
&= \frac{i}{2} \ln\frac{E\eta_1}{8I_p}. \quad (\eta_1 \to 0)
\end{aligned} \qquad (1.49)$$

And

$$Im S_{02} = \frac{1}{2} \ln\frac{E\eta_1}{8I_p}. \qquad (1.50)$$

Now we calculate the action S_1. The asymptotic expression of the atomic wave function at the distances $\eta < \eta_1$ has the form:

$$\phi_i^{(0)}(\eta) = \eta^{1/2} \exp(-\eta/2). \qquad (1.51)$$

Further, we compare Eq. (1.51) with the WKB expression

$$\phi_i^{(0)}(\eta) = \exp i S_1. \qquad (1.52)$$

Thus

$$iS_1 = \frac{1}{2} \ln\eta \quad (\eta < \eta_1 \to 0). \qquad (1.53)$$

So we can sum the action S

$$\begin{aligned}
ImS &= ImS_1 + ImS_{01} + ImS_{02} \\
&= \frac{1}{2} \ln\frac{1}{\eta_1} + \frac{(2I_p + p_\perp^2)^{\frac{3}{2}}}{3E} + \frac{1}{2} \ln\frac{E\eta_1}{8I_p}
\end{aligned}$$

$$= \frac{(2I_p + p_\perp^2)^{\frac{3}{2}}}{3E} + \frac{1}{2} \ln \frac{E}{4}$$

$$= \frac{(2I_p + p_\perp^2)^{\frac{3}{2}}}{3E} + \frac{1}{2} \ln \frac{E}{4} \tag{1.54}$$

so that the ionization rate of Hydrogen atom in constant electron field is

$$w = \frac{4}{E} \exp\left(-\frac{2(2I_p + p_\perp^2)^{\frac{3}{2}}}{3E}\right) \tag{1.55}$$

Expanding the result over the small momentum $p_\perp$, we have

$$w = \frac{4}{E} \exp\left(-\frac{2(2I_p)^{\frac{3}{2}}}{3E}\right) \exp\left(-\frac{(2I_p)^{1/2} p_\perp^2}{E}\right). \tag{1.56}$$

1.1.3 Extended to Hydrogen-Like Atoms

The above calculation can be extended to hydrogen-like atoms with energy $-I_p$, orbital quantum number l, and magnetic quantum number m [5]. The probability of ionization per unit time is described as:

$$
\begin{aligned}
w = {} & \left(\frac{3}{\pi^3}\right)^{\frac{1}{2}} \frac{(2l+1)(l+|m|)!}{(|m|)!(l-|m|)!} \\
& \times \left(\frac{e}{(n^{*2} - l^{*2})^{1/2}}\right)^{|m|+3/2} \left(\frac{n^* + l^*}{n^* - l^*}\right)^{l^*+1/2} \\
& \times \frac{Z^2}{n^{*3}} \left(\frac{4eZ^3}{En^{*3}(n^{*2} - l^{*2})^{1/2}}\right)^{2n^* - |m| - 3/2} \exp\left(-\frac{2Z^3}{3En^{*4}}\right), \tag{1.57}
\end{aligned}
$$

where E is the electric field strength, $n^* = 1/\sqrt{2I_p}$ is the effective principal quantum number, and l^* is the effective orbital number, defined as

$$l^* = n_0^* - 1,$$

where n_0^* is the effective principal quantum number of the ground state.

In the ultrahigh intensity laser fields, the tunnel ionization rates for hydrogen-like ions with charge Z can be obtained from a semiclassical solution of the three-dimensional Dirac equation [6]. In the relativistic system of units $c = m = \hbar = 1$, it reads as:

$$w = \frac{(eE)^{1-2I_p}}{2\sqrt{3}\xi\Gamma(2I_p+1)}\sqrt{\frac{3-\xi^2}{3+\xi^2}}(\frac{4\xi^3(3-\xi^2)^2}{\sqrt{3}(1+\xi^2)})^{2I_p}$$

$$\times \exp(6\mu \arcsin\frac{\xi}{\sqrt{3}} - \frac{2\sqrt{3}\xi^3}{eE(1+\xi^2)}), \tag{1.58}$$

where $\xi = \sqrt{1-(1/2)I_p(\sqrt{\varepsilon^2+8}-\varepsilon)}$, and $\mu = e^2Z$.

1.2 Classical Trajectory Model

When we consider the atom irradiated by intense laser fields, the laser fields are usually treated classically because the photon density is so high. On the other hand, since the laser field is strong compared to the Coulombic attraction, atomic ionization in this situation is largely related to the continuum and continuum–continuum transition and the classical description in this case provides an alternative way to explore the complex dynamics of atomic behavior. Compared to quantum mechanics treatment, the classical simulation is less time-consuming and intuitive picture providing.

1.2.1 Classical Trajectory Monte Carlo (CTMC) Approach

The classical hydrogen model consists of a nuclei of effectively infinite mass at rest at the origin of the coordinates and an electron. The electron is subject to the Coulomb field from the nuclei and also to a time-dependent plane electromagnetic wave whose Poynting vector is parallel to $+x$ axis. The linearized polarization directs to $+z$ axis. Neglecting the magnetic field and relativistic effects, the Hamiltonian can be written as

$$H = \frac{p^2}{2} - \frac{1}{r} + zF\cos\omega t, \tag{1.59}$$

where r and p are the position and the momentum of the electron; ω and F denote the laser frequency and the electric field of the laser respectively.

As the initial distribution of the ground state atom is microcanonical, we have

$$\rho_\mu(r,p) = \sigma(-I_p - H_0(r,p))/K, \tag{1.60}$$

where $H_0(r,p) = \frac{p^2}{2} - \frac{1}{r}$. A constant K is introduced to normalize the distribution.

Integrating Eq. 1.60, one obtains the momentum distribution $\rho(p)$ corresponding to $\rho_\mu(r,p)$,

$$\rho(p) = \frac{4\pi}{K} \int_0^\infty dr r^2 \sigma(-\frac{1}{2} - \frac{p^2}{2} + \frac{1}{r}) = \frac{8p_c^5}{\pi^2(p^2 + p_c^2)^4}, \tag{1.61}$$

where $p_c^2 = 2I_p$, I_p being the ionization energy of the atom. Interestingly, this classical microscopic distribution totally agrees with the distribution calculated from quantal ground wavefunction.

The microcanonical distribution can be obtained through the following processes [7]: assuming that the electron moves in the $y - z$ plane and solving Kepler's equation, we get the coordinates and momenta of the electron

$$\mathbf{r}_{2,y-z}^0 = \begin{pmatrix} 0 \\ a\sqrt{1-\varepsilon^2}\sin u \\ a(\cos u - \varepsilon) \end{pmatrix}, \quad \mathbf{p}_{2,y-z}^0 = \begin{pmatrix} 0 \\ b\sqrt{1-\varepsilon^2}\cos u/(1-\varepsilon\cos u) \\ -b\sin u/(1-\varepsilon\cos u) \end{pmatrix}. \tag{1.62}$$

where, $a = 1/2I_p$, $b = \sqrt{2I_p}$, ε is the eccentricity of the orbit with ε^2 randomly distributed in $[0, 1]$. u is the eccentric angle which has a complex distribution. In general, another new geometrical parameter is induced $\theta_n = u - \varepsilon\cos u$. It can be proved that θ_n is proportional to time, thus $0 \ll \theta_n \ll 2\pi$ corresponds to a periodic motion of an electron. If ε and θ_n are set, u can be obtained numerically. Finally, we get the initial coordinates and momenta of the electron. ϕ, θ, η are Euler angles, $(-\pi \ll \phi \ll \pi, -1 \ll \cos\theta \ll 1, -\pi \ll \eta \ll \pi)$. Performing a rotation of the above orbit by Euler angles, the electronic initialization is completed. $\mathbf{r}_2^0 = A\mathbf{r}_{2,y-z}^0$, $\mathbf{p}_2^0 = A\mathbf{p}_{2,y-z}^0$. The matrix of rotation is

$$A = \begin{pmatrix} -\sin\phi\sin\eta + \cos\phi\cos\theta\cos\eta & -\sin\phi\cos\eta - \cos\phi\cos\theta\sin\eta & \cos\phi\sin\theta \\ \cos\phi\sin\eta + \sin\phi\cos\theta\cos\eta & \cos\phi\cos\eta - \sin\phi\cos\theta\sin\eta & \sin\phi\sin\theta \\ -\sin\theta\cos\eta & \sin\theta\sin\eta & \cos\theta \end{pmatrix}. \tag{1.63}$$

In practical simulations, the integration program of Roung-Kuta can be used to solve the differential equations about position and momentum of each electron and trace individual time-resolved classical trajectory. As is well known, there is a singular point $r = 0$ in Eq. 1.59. Soft core (that is, taking $r + a$ instead of r when the collision occurs) may remove the singularity but cause uncontrollable deviation from the original system at the same time. This might be due to that the energy gain of each electron during ionization process depends strongly on the minimum distance between the electron and the nucleus. To our experience, the time transformation $dt/d\tau = r^3$ is an alternative way to regularize the motion of collision near origin in the dynamical evolution. To classify the behaviors of classical trajectories, it is convenient to introduce the compensated energy E_c advocated by Leopold and Percival [8, 9].

$$E_c = \frac{1}{2}(v_x^2 + v_y^2 + [v_z + (F/\omega)\sin\omega t]^2) - 1/r. \tag{1.64}$$

When an electron is ionized, the Coulomb potential is weak, and E_c is positive and nearly constant in time.

In the above CTMC method, it was assumed that the initial coordinates and momenta of bounded electron are uniformly distributed in phase space on a shell with single energy value, i.e., the microcanonical distribution. The obtained momentum distribution agrees with the quantum mechanical counterpart, while coordinate distribution deviates from quantum distribution. Some alternative ensemble distributions have been proposed [10]. It was found that if one scatters the electron over an energy range rather than a single value, one can generate a classical trajectory ensemble so that both the coordinate and momentum distributions agree with quantum distributions.

1.2.2 Classical Trajectory Monte Carlo with Tunneling Allowance (CTMC+T) Aproach

With the CTMC approach, through tracing individual classical trajectory of electron motion in combined Coulomb attraction potential and laser electromagnetic field and making statistics of trajectory ensemble, one can obtain the energy spectra and angular momentum distributions of the ionized electron and achieve insight into involved complex dynamics [11, 12].

The explicit expression of the classical ionization rate based on simplified 1D model was obtained [12] and compared to quantum KFR theory and Landau tunneling theory [1]. Near the threshold field the results of the Landau theory agree well with the numerically solving Schrödinger equation, while the Keldysh (strong-field approximation) theory underestimates the rates below or near the threshold field. The classical rates cannot be defined properly when the field strength is lower than the threshold value. In the regime of the above threshold field, the classical rates are in agreement with the quantum numerical computations.

Although useful classical calculations of strong-field ionization have been performed and classical arguments have frequently been invoked to explicate observed behaviors, the classical model that is rather accurate for high Rydberg states might be rough for the ground-state atom. In the regime of far below threshold or near threshold fields, the atomic ionization probability calculated from the classical approach is orders of magnitude too small. This defect of classical theory is mainly due to quantum tunneling, which is an important physical effect that has been totally ignored in the CTMC approach.

The tunneling effect can be compensated with allowance of tunneling whenever electron reaches classical turning point, where $p_{i,z} = 0$ and $z_i \varepsilon(t) < 0$, with a tunneling probability P_i^{tul} given by the WKB approximation [13]

$$P_i^{tul} = \exp\left[-2\sqrt{2} \int_{z_i^{in}}^{z_i^{out}} \sqrt{V(z_i) - V(z_i^{in})} dz_i \right]. \qquad (1.65)$$

Here, z_i^{in} and z_i^{out} are the two roots ($\left|z_i^{out}\right| > \left|z_i^{in}\right|$) of the equation for z_i, $V(z_i) = -2/r_i + z_i \varepsilon(t) = -2/r_i^{in} + z_i^{in} \varepsilon(t)$.

Fig. 1.2 Laser intensity dependence of ionization rate. Squares and empty squares represent CTMC+T result and pure CTMC result, respectively. Triangles are from directly solving time-dependent schrödinger equation [14]. The dashed line represents the results of ADK formula in Eq. (1.57)

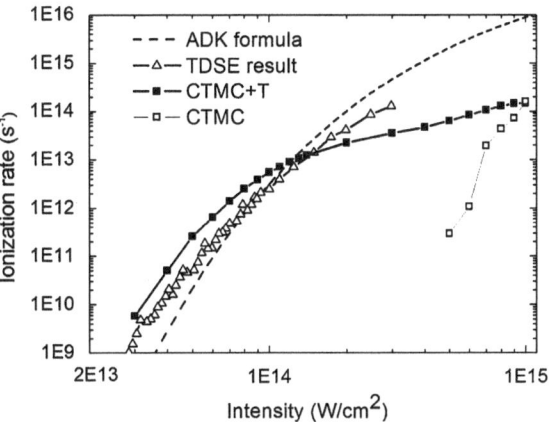

We calculate the laser intensity dependence of ionization rates for hydrogen atom using CTMC+T and CTMC models for comparison. Figure 1.2 shows that the pure classical CTMC fails to spell out the correct ionization rate in the regime below the threshold field. While, with allowance of tunneling, CTMC+T model calculations give a reasonable result of ionization rates for a wide range of laser intensity that fairly agrees with the results from directly solving the time-dependent Schrödinger equation. In the larger field strength regime, the ADK theory seems to overestimate ionization yields, while below $10^{14} W/cm^2$ it predicts smaller ionization rates compared to CTMC+T and quantum model calculations.

References

1. Landau, L.D., Lifshitz, E.M.: Quantum Mechanics, 3rd edn. Pergamon, Oxford (1977)
2. Perelomov, A.M., Popov, V.S. and Teren'ev, V.M.: Zh. Eksp. Teor. Fiz. 52, 514 (1967) [Sov. Phys. JETP 25, 336 (1967)].
3. Delone, N.B., Krainov, V.P.: J. Opt. Soc. Am. B **8**, 1207 (1991)
4. Popov, V.S.: Physics - Uspekhi **47**(9), 855–885 (2004)
5. Ammosov, M.V., Delone, N.B., Krainov, V.P.: Sov. Phys. JETP **64**, 1191 (1986)
6. Milosevic, N., Krainov, V.P., Brabec, T.: Phys. Rev. Lett. **89**, 193001 (2002)
7. Cohen, James S.: Phys. Rev. A **26**, 3008 (1982)
8. Leopold, J.G., Percival, I.C.: J. Phys. B **12**, 709 (1979)
9. Abrines, R., Percival, I.C.: Proc. Phys. Soc. London **88**, 861 (1966)
10. Cohen, J.S.: J. Phys. B: At. Mol. Phys. **18**, 1759–1769 (1985)
11. Liu, Jie, Shi-gang, Chen, De-hai, Bao: Acta Physica Sinica (overseas edition) **4**(12), 881–887 (1995)
12. Liu, Jie, Chen, Shi-gang, Bao, De-hai: Commun. Theor. Phys. **25**, 129–134 (1996)
13. Cohen, J.S.: Phys. Rev. A **64**, 043412 (2001)
14. K.C. Kulander, K.J. Schafer, and J.L. Krause, Atoms in Intense Laser Fields, edited by M. Gavrila (Academic, San Diego, 1992), p. 247.

Chapter 2
Single Ionization in Strong Laser Fields

Abstract In this chapter, we discuss single ionization of hydrogen-like atom irradiated by an intense laser field based on the semiclassical simulation of classical trajectory ensemble. We present the interesting energy spectrum and irregular angular distribution of photoelectrons, and discuss partial atomic stabilization and associated atomic survival window. Finally, we show how chaotic trajectories play role in the rescattering process of tunneled electron.

2.1 Semiclassical Model for Single Ionization

As Keldysh parameter $\gamma \equiv (I_p/2U_p)^{1/2} \ll 1$, $\hbar\omega/2U_p \ll 1$ and field strength $E < E_{th}$, tunneling ionization occurs [1, 2]. Here, I_p is ionization potential, ω is field frequency, $U_p = \frac{e^2 E^2}{4m_e\omega^2}$ is ponderomotive potential, and E_{th} is the threshold field above which atomic ionization goes into over-the-barrier regime. The Keldysh parameter γ is an important parameter in strong-field atomic ionization. It can be used to demarcate the boundary between tunneling ionization (i.e., $\gamma < 1$) and multiphoton ionization $\gamma > 1$. The physical meaning of the Keldysh parameter is the ratio between field frequency ω and tunneling frequency ω_t. Tunneling frequency is inversely proportional to tunneling time, which is estimated as the tunneling distance divided by the electron velocity, i.e., $(I_p/E)/\sqrt{2I_p}$ [3].

In the tunneling regime ($\gamma < 1$), atomic ionization consists two essential physical processes, i.e., electron tunnels through the Coulomb field that has been dramatically suppressed by laser field, and the released electron is driven by laser filed to scatter with its parent ion. The latter is termed as rescattering process [4]. The first step is of intrinsic quantum property. As was discussed in the first chapter, the tunneling of a hydrogen-like atom is preferably discussed in parabolic coordinates. Because the laser frequency is small compared to the tunneling frequency, the field can be regarded as quasistatic, then the reduced 1D effective Schrödinger equation takes following form [5],

$$\frac{d^2\phi}{d\eta^2} + (\frac{I_p}{2} + \frac{1}{2\eta} + \frac{1}{4\eta^2} + \frac{1}{4}E(t_0)\eta)\phi = 0. \tag{2.1}$$

in which $E(t_0)$ is the instantaneous electric field. The above equation has the form of one-dimensional Schrödinger equation with the potential $U_{eff} = -\frac{1}{4\eta} - \frac{1}{8\eta^2} - \frac{E(t_0)\eta}{8}$ and the effective ionization energy becomes $K = -\frac{I_p}{4}$. The turning point η_0 at the outer edge of the distorted Coulomb potential, is determined by $U_{eff}(\eta_0) = K$, which represents the location where the tunneling electron borns. There exists a threshold field E_{th}, when $E > E_{th}$, the turning point becomes complex, indicating the transition to over-the-barrier regime. The above discussions provide the location configuration of the electron that releases at time t_0. For the linearly polarized laser, when the instantaneous field is along z axis, the initial coordinates of the tunneling electron will be $x_0 = y_0 = 0, z_0 = -\eta_0/2$. According to the tunneling theory, the tunneling electron will have nonzero transversal velocity and zero longitudinal velocity, i.e., $v_z(t_0) = 0, v_x(t_0) = v_{x0}, v_y(t_0) = v_{y0}$

After tunneling, the electron's motion is described by rescattering dynamics that mainly involves the continuum states and has distinct classical feature. We, therefore, treat it within the framework of classical mechanics. For a hydrogen-like atom, with ignoring the magnetic component of the laser field, the Newtonian equations of the valence electron reads,

$$\ddot{\mathbf{r}}_\mathbf{e} = -\frac{\mathbf{r_e}}{r_e^3} - f(t)\cos\omega t\mathbf{e_z}. \tag{2.2}$$

Here, $f(t)$ is the laser pulse and ion is set frozen at origin.

Tunneling rate corresponding to the instantaneous field $E(t_0)$ should be assigned to each classical trajectory. That is, $w(t_0, v_\perp) = w(0)w(1)$, where

$$w(0) = \left(\frac{E(t_0)}{4}\right)\left(\frac{4\kappa^4}{E(t_0)}\right)^{2/\kappa} e^{-\frac{2\kappa^3}{3E(t_0)}}, w(1) = \frac{\kappa}{\pi E(t_0)}e^{-\kappa((v_{x0})^2+(v_{y0})^2)/E(t_0)}, \tag{2.3}$$

in which the parameter $\kappa = \sqrt{2I_p}$.

With this, we associate each trajectory j (tunneling at t_0) with a weight $w_j = w(t_0, v_\perp)$. We then can calculate a physical observable quantity O through weighed average scheme as

$$\langle O \rangle = \sum_j O_j w_j / \sum_j w_j, \tag{2.4}$$

here O_j represents the result from an individual j trajectory. In practical simulation, the tunneling time or field phase ωt_0 and transversal velocity $v_\perp$ are scattered in proper ranges in the spirit of Monte-Carlo method and trajectory number is large enough so that the results converge [6].

The above discussions can be readily extended to general non-relativistic laser field of form, $f(t)(\cos\omega t\mathbf{e_x} + \chi\sin\omega t\mathbf{e_y})$ in which $x - y$ is the polarization plane and χ is elliptic polarization eccentricity. $\chi = 0$ corresponds to linearly polarized and $\chi = \pm 1$ corresponds to circularly polarized laser.

For the general elliptically polarized laser field, we align the z axis along the laser propagation direction, and assume the field to be quasistatic. So we can use the results developed above. For every initial time t_0, we can calculate the electron tunneling point which is located in the reverse direction of the instantaneous electric field with a distance of $\left| \frac{\eta_0}{2} \right|$ away from the core. We also assign the electron the transverse velocity $v_\perp$ perpendicular to instantaneous field, so the velocity components of the tunneled electron are

$$v_{x0}^e = -v_\perp \sin \beta \sin \theta, \, v_{y0}^e = v_\perp \sin \beta \cos \theta$$
$$v_{z0}^e = v_\perp \cos \beta \tag{2.5}$$

Here, θ is the angle between the instantaneous electric field and positive x axis, and β the angle of $v_\perp$ from the positive z direction. We attribute a rate $w(t_0, v_\perp)$ to each trajectory. We then can calculate the average physical quantities using the previously described method.

2.2 Plateau in Above-Threshold-Ionization (ATI) Spectrum and Irregular Photoelectron Angular Distribution (PAD)

One of most important findings in recent experiments on atomic ionization is the plateau formed by high-order ATI peaks, which halt and even temporarily reverse the decrease of the heights of the peaks with increasing order [7–9]. Moreover, the photoelectron angular distribution (PAD) in the transition region at the onset of the plateau seems unusual: while the angular distributions, both below and well above this region, are strongly concentrated in the field direction, additional "side lobes" at angles between $\pi/6$ and $\pi/4$ with respect to the field direction develop in the transition region [7–9]. The plateau structure is universal and emerges for noble atoms. It is obviously out of the traditional perturbation theory. To better understand the underlying physics and obtain intuitive picture, we use our developed semiclassical model to investigate the ionization dynamics of hydrogen atom in an intense laser field [6].

In the simulation, Rung-Kutta self-adapted algorithm is exploited to solve Newtonian differential equations. More than 10^5 trajectories are traced. In fig. 2.1a, we demonstrate the ATI spectra and the total angular distribution of the emitted electrons with respect to the angle θ in positive z direction, i.e., at the detector, calculated from our model. Compared to "simple-man" model that predicts a sharply decreasing ATI spectra curve with maximum photoelectron energy of $2U_p$ [10–12], our model calculation shows clearly that rescattering increases the fraction of the hot electrons. This is due to that an electron has a higher probability of staying in the vicinity of the nucleus and then absorbing more photons. In particular, the ATI spectrum exhibits a sharply decreasing slope followed by a plateau (the transition region) and again

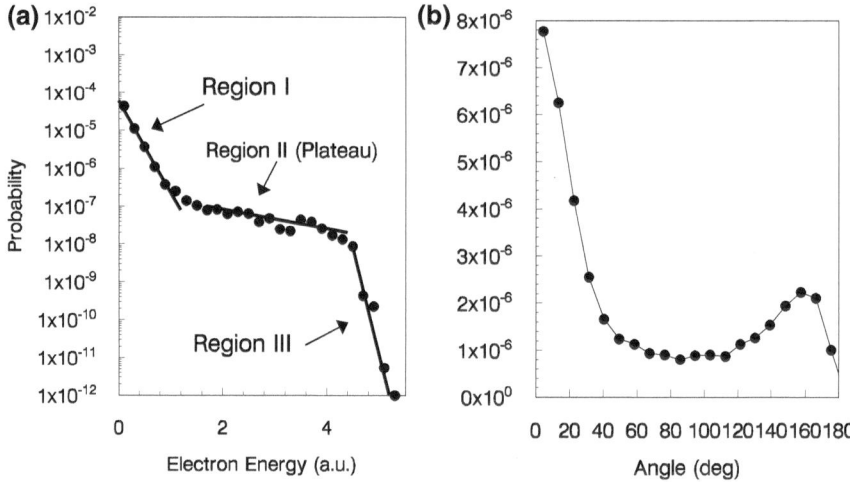

Fig. 2.1 **a** ATI spectrum and **b** total PAD calculated from our model. Reprinted with the permission from Ref [6]. Copyright 1997

a sharply decreasing slope. The height of the plateau is three orders of magnitude below the maximum of the spectrum and its width is about $5U_p$. This phenomenon is qualitatively in agreement with existing experiments and other quantum theories [7–9, 13, 14]. The distinctive feature is the plateau's extension about $5U_p$ width to rather high energy before abruptly declining at about $8U_p$, is much closer to the recent experiment [15] which is performed in the strong tunneling limit. The total angular distribution shown in Fig. 2.1b contains only the electron initiated in the phase interval $[-\pi/2, \pi/2]$. The angular distribution of the electrons originated in $[\pi/2, 3\pi/2]$ will be the mirror image with respect to 90°, so the sum of the two contributions will show a main concentration in the field direction.

Furthermore, we calculate the statistics on the angular distribution of photoelectrons in three different energy regions shown in Fig. 2.2. The most striking feature of the plots is the existence of a slight slope up to 40° followed by a sharp cut-off, i.e., no photoelectrons in the transition region emitted at angles much larger than 40° (see Fig. 2.2b). This remarkable phenomenon corroborates the data of Paulus *et al* [16], which we think is due purely to tunneling ionization. In contrast, in the mixing regime where multiphoton ionization becomes significant, the angular distributions show no such cut-off and there appears to be emission even at 90°. Considering the rescattering effects from a simple classical model, Paulus et al find a peak at 30° of the angular distribution for the electrons in the transition region. However, the emission of the photoelectrons in the direction of the laser electric field is largely underestimated [17]. In our model, this probability increases greatly because a complete Newtonian equation is used to simulate the evolution of an electron after tunneling. Furthermore, the PAD in the plateau regime exhibits an additional peak which is called a side-lobe phenomenon observed in experiment. One example is shown in

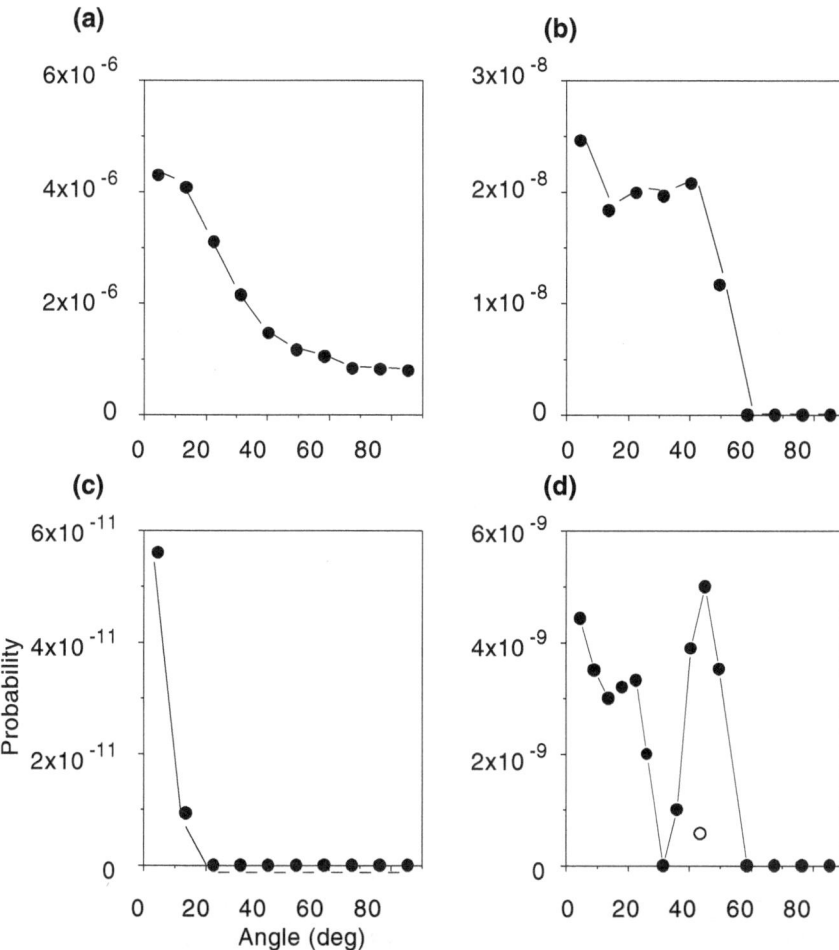

Fig. 2.2 PADs for different energy regions. **a,b** and **c** correspond to the region I,II, and III in Fig. 2.1 a, respectively, **d** shows the PAD at energy around $4U_p$, where the side lobe near 40° is very obvious. Reprinted with the permission from Ref [6]. Copyright 1997

Fig. 2.2d for $4U_p \pm 0.1U_p$. There the side lobe occurs at about 40°. The sum of the side lobes for photoelectron in the plateau region is responsible for the slight slope followed by a sharp cut-off observed in Fig. 2.2b. Our results also show that the angular distributions both below and well above the plateau region are strongly concentrated in the field direction. Detailed investigations show that, aside from these electrons which directly drift away without returning to the core, most electrons in the region I experience forward scattering, while the electrons in the region III are backscattered by almost 180°. However, in the plateau region, forward scattering and backward scattering are equally important and seem to have equal probability.

2.3 Chaotic Behavior in Rescattering Process

Classical trajectory description of electron's motion provides an intuitive picture of the rescattering dynamics in strong-field atomic ionization. In this section, taking the hydrogen for example, we show that chaotic trajectory is evident and responsible for the observed unusual ATI spectrum and PADs. The rescattering process of the electrons after tunneling can be described by the Newtonian Eq. (2.2). ATI spectra and PADs can be obtained by calculating statistics on an ensemble of trajectories corresponding to different initial field phase and perpendicular velocity. To investigate the detailed dynamical mechanism underlying these unusual ATI spectra and PADs, we calculate the initial phase($\phi_0 = \omega t_0$) dependence of ATI energy and emission angle for two typical initial perpendicular velocities (Fig. 2.3). As the initial transversal velocity is relatively large (Fig. 2.3a and 2.3b), we have smooth phase dependence ATI energy and emission angle which imply regular motion of classical trajectories. The phase dependence spectrum (see Fig. 2.3a) is much similar to that of

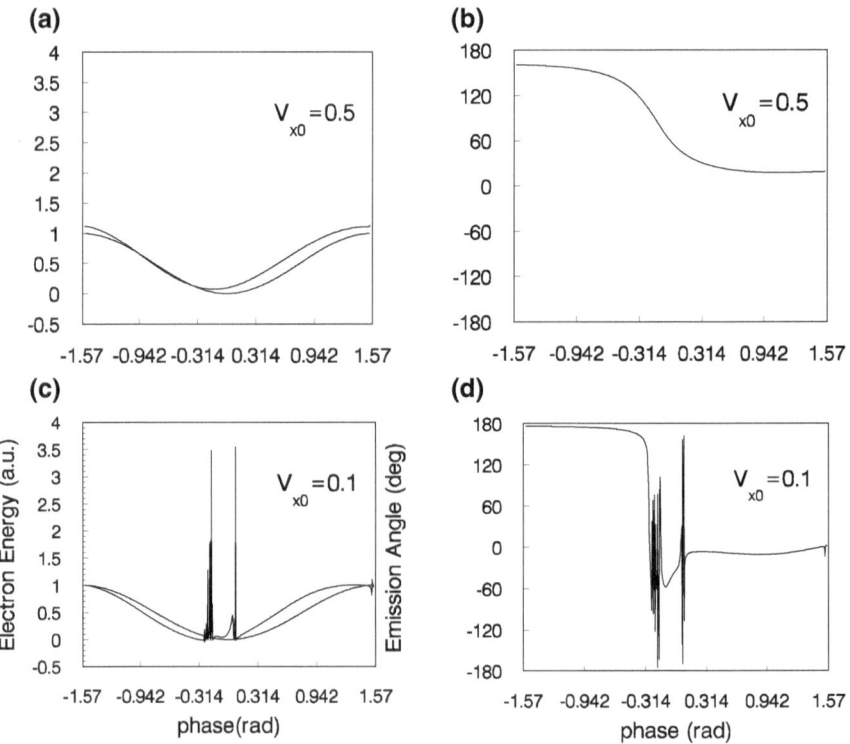

Fig. 2.3 The *solid lines* denote the phase dependence ATI energy spectra and PADs calculated from our model for two Different initial perpendicular velocities. The *dashed lines* represent the phase dependence ATI energy calculated from the simple-man model for comparison. Reprinted with the permission from Ref [6]. Copyright 1997

the simple-man model, where the effect of the Coulomb potential is ignored so that all motions of the classical electrons are regular. The energy of the ionized electron is less than $2U_p$. This case corresponds to a sharply decreasing ATI spectrum and PAD. Therefore, these trajectories have very small contribution to the unusual distributions in ATI data.

For case of relatively smaller perpendicular velocity (Fig. 2.3c and 2.3d), things are quite different. The phase dependence spectrum (see Fig. 2.3c) deviates dramatically from the prediction of simple-man model. The dependence of ATI energy and the emission angle on the initial phases is poorly resolved in a region near zero point where many prominent peaks are observed. Successive magnifications (Fig. 2.4) of the unresolved regions show that any arbitrarily small change in the initial phase may result in a substantial change of the final electron energy and emission angle. Multiple returns and even infinitely long time trapping can occur in this region. This

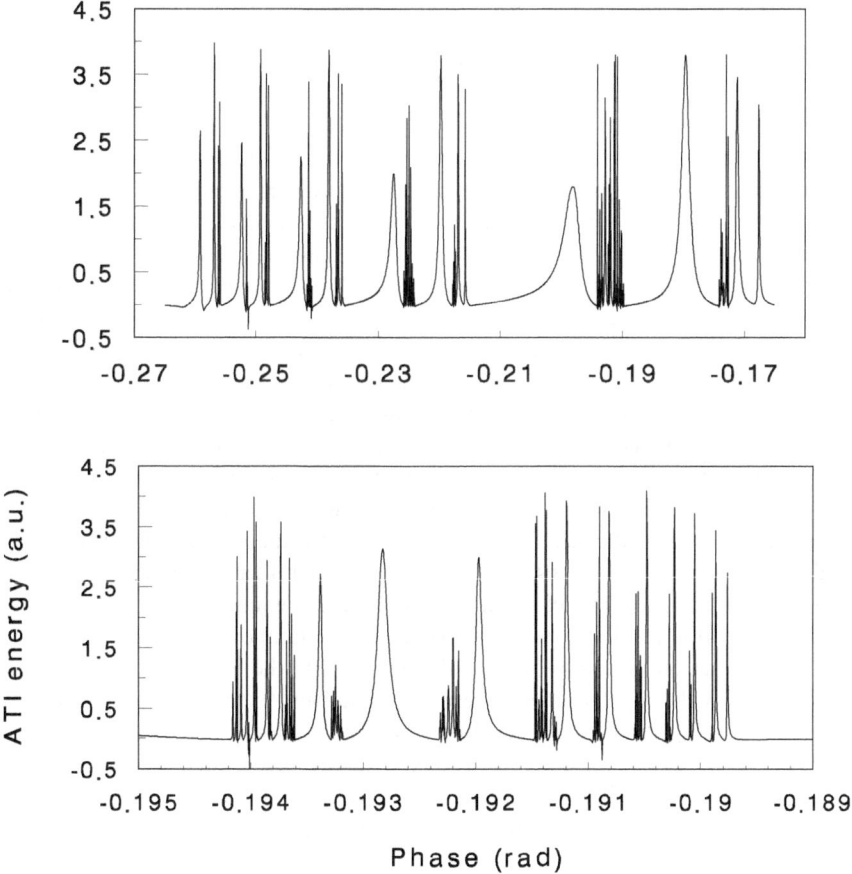

Fig. 2.4 Successive magnifications of Fig. 2.3c. Reprinted with the permission from Ref [6]. Copyright 1997

is the characteristic of chaotic behavior, namely, *sensitive dependence on the initial condition*, which has been observed in many chaotic scattering models [19, 21]. In the unresolved regions, some chaotic trajectories of higher probability can produce relative high ATI energy (larger than $2U_p$). These chaotic trajectories are responsible for the unusual structure–plateau in ATI spectra. Moreover, the emission angles of these trajectories are also strange (Fig. 2.3d), then leading to the unusual angular distribution in the transition region.

2.4 Partial Atomic Stabilization in Strong-field Tunneling Ionization

Atomic stabilization (or ionization suppression), that the total ionization yield decreases or at least cease to increase as the laser intensity increases, has been studied in superintense high-frequency fields theoretically for decades (see, e.g., [20] for a review,) and has led to a wealth of profound and intrigue concepts in strong-field physics [22–26]. In the Kramers-Henneberger (KH) frame, i.e., the moving coordinate frame of a free electron responding to a monochromatic laser field, the ground state wave function of the atom splits into two non-overlapping peaks and the atom becomes stabilized against ionization when the laser frequency is higher than the bound state frequency of the atom [27–30]. Recent theoretical investigations further reveal that this concept is not exclusively associated with high frequencies, as widely assumed [31, 32]. On the other hand, in an intense low-frequency (e.g., in the infrared regime) laser field, the tunneling limit of multiphoton ionization is more appropriately described by the tunneling theory [33]. Along this, an effect of ionization suppression associated with tunneling was predicted at some specific field strengths [34]. The controversial issue in the deduction, however, has been argued recently [35], in which the dependence of the ionization rate on the laser field is found to be monotonic. In contrast to the high-frequency multiphoton ionization, the low-frequency atomic stabilization in the tunneling regime is a more subtle and intriguing question. In a recent experiment [48], it was observed that, as the intensity of the field increases, the relative contribution of low-energy photoelectrons to the total ionization yield decreases, i.e., partial atomic stabilization. In this section, we apply our semiclassical model to get insight into the phenomenon.

The simulation results of two-dimensional momentum distributions of $(P_z, P_\perp)$ for Xe atoms at the intensity of 6×10^{13} W/cm^2 and the wavelength of 1320 nm result agrees with the experimental results qualitatively except for the interference patterns. In particular, it reproduces the phenomenon of local ionization suppression in PADs at the origin. In order to trace the suppressed events, we illustrate the energy distribution of all electrons that tunnel from the first half of the laser cycle, regardless of whether the final energy is positive or negative, in Fig. 2.5a. It exhibits that, after the laser pulse there are a large number of tunneled electrons with the final negative energies within $(-0.01, 0)a.u.$, which means that those electrons are finally bounded by the atomic potential. The binding energies of those Rydberg states

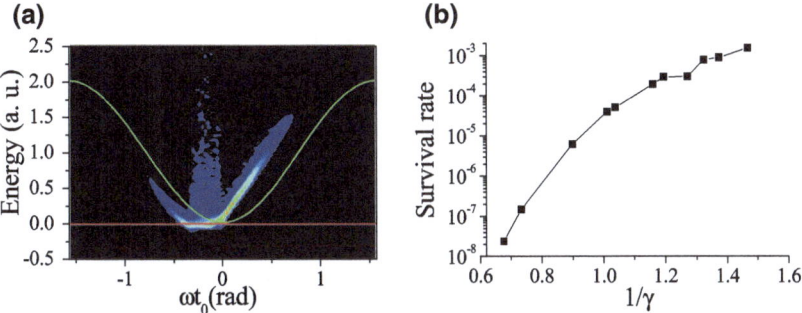

Fig. 2.5 Simulation results for Xe at the intensity of 6×10^{13} W/cm^2 at 1300 nm. **a** The energy distribution of all electrons vs. the initial phase. The *green curve* shows the prediction of the simple-man model. **b** The survival rate with respect to the inverse Keldysh parameter. Reprinted with the permission from Ref [48]. Copyright 2012, American Physical Societ

($E < 0.01$ a.u.) are much smaller than the photon energy of the low-frequency light (for 1320 nm, the photon energy is 0.04 a.u.). It has been found experimentally that a large number of excited neutral atoms can survive in strong laser fields [37]. The fact that a substantial part of the tunneled electrons end up in the bound states should affect the momentum spectrum of ionized electrons, particularly the low-energy part. Moreover, the survival rate increases with the inverse Keldysh parameter as presented in Fig. 2.5b. This is consistent with the experimental observation that the relative yields of near-threshold momentum electrons are more suppressed in deep tunneling regime. As shown in Fig. 2.5a, in the presence of the Coulomb field, the energies of the electrons released at the rising front are depressed and the energies of the electrons released at the descending front are enhanced, as compared with the prediction by the simple-man model [10–12] (the solid green curve in Fig. 2.5a). Some electrons tunneled from the phase region slightly before the field maximum can achieve much higher energy than 2Up, which are resulted from the electron chaotic motions [6, 38].

Figures 2.6a and b show two typical electron ionization trajectories associated with the survival atoms. The electrons born with a certain field phase and transverse velocity at the rising front of laser pulse can finally be launched into the elliptical orbits that have the negative energy. In the first optical cycle after tunneling, the electron obtains or releases energy depending on the instantaneous field phases, and then is pumped into the Rydberg elliptical orbits. The difference between those two typical trajectories is that, the electron in Fig. 2.6a is ejected directly into an elliptical orbit without collision with the atomic ion, while the tunneled electron in Fig. 2.6b experiences the hard collision with the core during its launching process. Our statistics indicate that the first type orbits constitute 90 % and the second type orbits contribute the residual 10 % of the total unionized electrons, respectively. We illustrate the statistic analysis on the semi-major axis and the eccentricity of the elliptic orbits in Fig. 2.6c and 2.6d. Those electrons with collisions prefer to move in the elliptic orbits with the smaller semi-major axes and larger eccentricities.

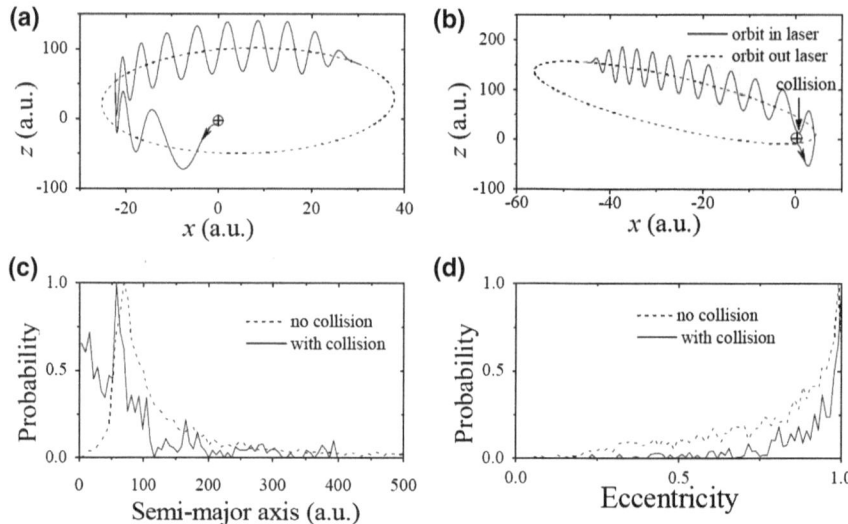

Fig. 2.6 **a** and **b** The typical trajectories of energy less than zero after tunneling and energy less than zero at the end of laser pulse, respectively. The first type trajectory represents the tunneled electron ejected into the elliptical orbits without collisions, and the second one experiences collisions with nucleus. **c** and **d** Statistic analysis on the semi-major axis and eccentricity distributions corresponding to those two typical trajectories, respectively. Reprinted with the permission from Ref [48]. Copyright2012, American Physical Society

The energy of an electron in the elliptic obit is related with the semi-major axis a, which is given by $E = -1/2a$. One can find that the distribution of semi-major axis mostly falls into the regime $1/2a << \omega$, i.e., the binding energies of the Rybderg states are much smaller than the photon energy. Because the classical elliptic orbit frequencies ($1/a^{3/2}$) are much smaller than the laser frequency, the fast oscillating motion driven by laser field can be safely averaged out and the electrons will finally remain on the elliptical orbits (see Fig. 2.6a and 2.6b). This is analogous to the stabilization condition for the Rydberg atoms in the low-frequency light field [27–30]. However, in our case, the electrons are released from the ground state through tunneling rather than prepared in the Rydberg states directly, the stringent atomic stabilization can not be observed.

2.5 Survival Window for Atomic Tunneling Ionization with Elliptically Polarized Laser Fields

Form the above discussion, we see surprisingly that, even experiencing intense irradiation from a laser, some atoms can still survive. The yields of neutral atoms and creating Rydberg atoms in laser field are of great importance in both fundamental and applied physics, such as acceleration of neutral atoms or molecules [39, 40] controlled collision [41–44], atomic nanofabrication[45, 46] and atom optics [47]. It is no doubt that Coulomb force dominates the evolution of the electron which tunneled without ionization [49]. In this section, we extend to investigate neutral atoms survival rate for elliptical laser fields based on our semiclassical quasi-static model with addressing the Coulomb effect [36].

In the model, the electric field rotates clockwise or counterclockwise as it propagates along z-axis. The tip of the electric field vector describes an ellipse in x-y plane. At time t_0, electron tunnels out from the nucleus parallel to the instantaneous electric field direction with zero initial parallel velocity. The initial tunneling position along the laser polarization direction can be derived from the Landau's effective potential theory [5]. Besides, the electron also has an initial transverse velocity $v_\perp$ perpendicular to the instantaneous electric field, and $v_\perp$ satisfies Gussian-like distribution. Each electron trajectory is weighted by the ADK ionization rate $w(\chi, t_0, v_\perp) = w(0)w(v_\perp)$ [15]. $w(v_\perp) = \frac{2\sqrt{2I_p}v_\perp}{\varepsilon_{t_0}} \exp\left(-\frac{\sqrt{2I_p}v_\perp^2}{\varepsilon_{t_0}}\right)$ is the distribution of initial transverse velocity $v_\perp$, and $w(0) = \varepsilon_{t_0}^{(1-2/\sqrt{2I_p})} \exp\left(-\frac{2(\sqrt{2I_p})^3}{3\varepsilon_{t_0}}\right)$ depends on the instantaneous field strength ε_{t_0} at the time when the electron releases and the ionization potential I_p. The evolution of the tunneled electron is governed by Newton's equations of motion, $\frac{d^2\mathbf{r}}{dt^2} = -\frac{\mathbf{r}}{r^3} - \boldsymbol{\varepsilon}$. The electric field of the laser is given by

$$\boldsymbol{\varepsilon}(t) = \frac{\varepsilon_0 f(t)}{\sqrt{1+\chi^2}}[\cos(\omega t)\mathbf{e}_x + \chi \sin(\omega t)\mathbf{e}_y], \tag{2.6}$$

where ε_0 and ω are the amplitude and frequency of the laser field,respectively. χ is the laser ellipticity. The envelope function $f(t)$ is the slowly-varying pulse envelope: $f(t) = 1$ for $t \leq 8T$, and adiabatically ramped off within three laser cycles, namely $f(t) = \cos^2 \frac{(t-8T)\pi}{6T}$ for $8T < t \leq 11T$. Here, $T = 2\pi/\omega$ is the oscillating period of the laser field. In our calculation the wave length is $\lambda = 800nm$ ($\omega = 0.056a.u.$).From the Landau's effective potential theory [5] and considering the instantaneous direction of field, the initial position can be given by, $x_0 = -\frac{\eta_0}{2}\cos\theta$, $y_0 = -\frac{\eta_0}{2}\sin\theta$, $z_0 = 0$, θ is the angle between the direction of the electric field and x-axis, in which $\eta_0 = \frac{I_p + \sqrt{I_p^2 - 2\varepsilon_{t_0}}}{\varepsilon_{t_0}}$, and $\varepsilon_{t_0} = \frac{\varepsilon_0\sqrt{\cos^2(\omega t_0)+\chi^2\sin^2(\omega t_0)}}{\sqrt{1+\chi^2}}$. The initial velocity is then $v_{x0} = -v_\perp \sin\beta \sin\theta$, $v_{y0} = v_\perp \sin\beta \cos\theta$, $v_{z0} = v_\perp \cos\beta$ with the distribution $w(v_\perp)$. The angle between the transverse velocity and the z-axis is β.With the model we calculate yields of neutral excited He* atoms and Mg* atoms for different ellipticities. The yields of neutral atoms are normalized as following $$W_{atom*} = \frac{\sum_{E_f<0} w(\chi,t_0,V_\perp)}{Max \sum_{E_f<0} w(\chi,t_0,V_\perp)},$$ E_f is the final energy of the tunneled electron.

The normalized yields of neutral excited He* atoms and Mg* atoms for different ellipticities are plotted in upper panels(Fig. 2.7): the left is for He* and the right is for Mg*, respectively. The results for helium obtained by our model are in good agreement with the experimental data, and consist with the prediction of SFA model for which the neutral atom yield is a Gaussian distribution as a function of ellipticity χ with a standard deviation $\sigma_0 = \sqrt{\frac{3}{3+\gamma^2}} \frac{\omega}{\sqrt{2\varepsilon_0}(2I_p)^{1/4}}$, where γ is the Keldysh parameter $\gamma = \frac{\omega\sqrt{2I_p}}{\varepsilon_0}$ [49]. At $\chi = 0$, the yield of He* is maximum. Near $\chi = 0.3$, the yield of survival helium decreases to zero. Different from helium, the yield of Mg* decreases slowly with increasing ellipticity. And even for circular polarization, the survival yield does not drop down to zero. Compared with the results from SFA model, the distribution of neutral atom yield w_{Mg^*} is much wider. This fact implies that the Coulomb effect may play an important role in the neutral atom surviving process for atom with smaller ionization potential.

Coulomb effects are twofold: the Coulomb potential of the tunneled electron at birth time and the Coulomb scattering effect during recollision process. Considering the Coulomb potential, the energy of tunneled electrons at the birth time can be expressed by $E_0 = \frac{1}{2}(v_0 + A(\omega t_0))^2 - \frac{1}{r_0}$, where r_0 is the exit point from tunneling and A is vector potential. Then, the tunneled electron will be accelerated in the consequent scattering processes mediated by the Coulomb and laser fields. The energy gain can be expressed $\Delta E = -\int_{t_0}^{t_{final}} \frac{xA_x + yA_y}{r^3} dt$ The electron orbit $\mathbf{r}(t) = (x(t), y(t), z(t))$ can be obtained under SFA by solving Newton equations $\frac{d^2\mathbf{r}}{dt^2} = -\mathbf{\varepsilon}$ with initial condition $\mathbf{r}_0 = (-\frac{\eta_0}{2}\cos\theta, -\frac{\eta_0}{2}\sin\theta, 0)$. Under this approximation, the final energy of electron is $E_f = E_0 + \Delta E$. When the tunneled electrons are released in a certain window of initial field phase ωt_0 and transverse velocity $v_\perp$, the final energy $E_f < 0$. We call this window as the survival window. The boundary line $E_0 + \Delta E = 0$ of survival window are plotted in Fig. 2.7c and 2.7d (labeled by red lines) for He and Mg, respectively.

2.6 Classical Trajectory Interference

In the semiclassical model, it is assumed that wavepacket propagation in the post-tunneling process can be well described within classical mechanics framework. This consideration comes from the classical-quantum correspondence principle for continuum quantum states. Sometimes, interference between quantum states cannot be ignored [50]. In this section, we extend our semiclassical model to include the interference between classical scattering trajectories.

We start with strong field approximation, where only atomic ionization potential I_p is concerned. The transition amplitude between the bound state to the continuum state of asymptotic momentum $\mathbf{p}$, using the saddle point approximation, is a coherent superposition of contribution from all relevant quantum orbits (indexed by α) that lead to the asymptotic momentum $\mathbf{p}$, i.e.,

$$M_{\mathbf{p}}^{SFA} \sim \sum_{\alpha} C_{\mathbf{p}}(t_s^{(\alpha)}) \exp(i W_0(t_s^{(\alpha)})) \tag{2.7}$$

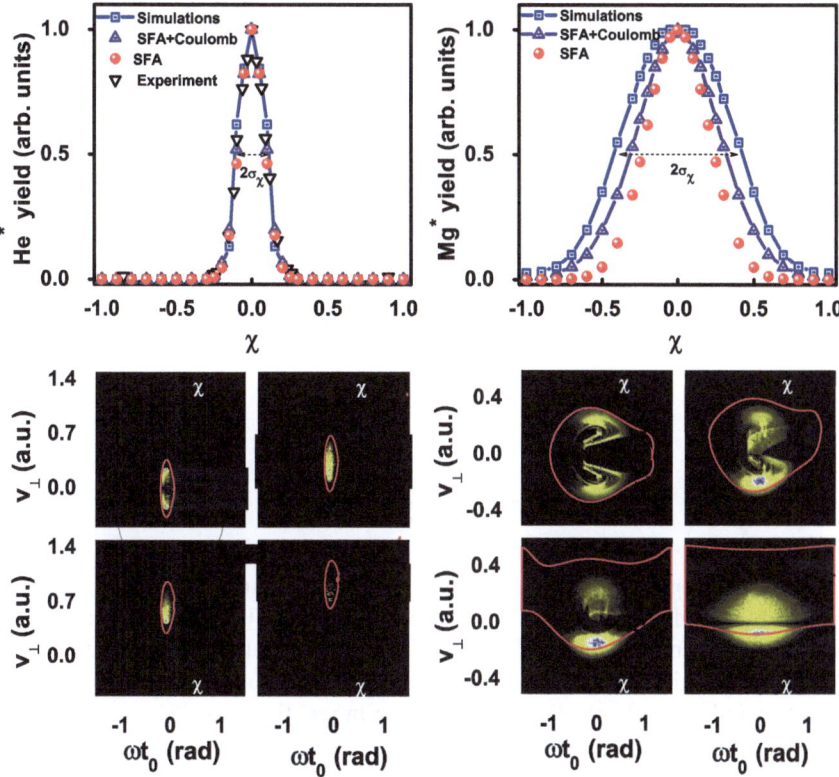

Fig. 2.7 *Upper panels* Yield of survival He* atoms (*left*) and Mg* atoms (*right*) versus elliptic-ity.*Black triangles* are experimental data from Ref. [18], *blue squares* are numerical simulations, *purple triangles* are obtained by SFA with Coulomb correction and the *red balls* are pure SFA results. The parameters used for He* atom is $I = 1\text{PW}/\text{cm}^2$, $\omega = 0.056$a.u. and for Mg* atom is $I = 0.0116\text{PW}/\text{cm}^2$, $\omega = 0.056$a.u.. *Lower pannels* Distribution of initial transverse velocity and tunneling phase for surviving He* atoms (*left*) and Mg* atoms (*right*) at different ellipticities. The bright regimes are numerical results and *red curves* indicate the boundary for unionized window predicted by our theoretical analysis. Reprinted with the permission from Ref [36]. Copyright 2013, American Physical Society

$C_\mathbf{p}(t_s^{(\alpha)})$ is a prefactor whose form can be found in [51, 52]. And the action

$$W_0(t_s) = -\int_{t_s}^{\infty} \left[\frac{1}{2}\mathbf{v}^2(t) + I_p\right] dt, \qquad (2.8)$$

where the velocity

$$\mathbf{v}(t) = \mathbf{p} + \mathbf{A}(t) \qquad (2.9)$$

corresponds to the motion of free electron in a laser field with the vector potential of $\mathbf{A}(t)$ in dipole approximation. The formula (2.8) can be interpreted as the motion which starts from complex time t_s, and satisfy the equation of motion,

$$\frac{d\mathbf{v}}{dt} = -\mathbf{E}(t), \tag{2.10}$$

$$\mathbf{E}(t) = -\frac{\partial}{\partial t}\mathbf{A}(t). \tag{2.11}$$

The releasing time is determined by the saddle point equation

$$\frac{\partial}{\partial t}W_0(t)|_{t_s} = 0, \tag{2.12}$$

i.e.

$$\frac{1}{2}[\mathbf{p} + \mathbf{A}(t_s)]^2 = -I_p \tag{2.13}$$

The integration from t_s to ∞ in Eq. (2.8) can be split into two portions, i.e., W^{sub} from t_s to t_r and W^{re} from t_r to ∞, where $t_r = Ret_s$. Provided the integrand is analytic, the result for the total action W is independent of the integration path. Before t_r the trajectory motion is imaginary indicating tunneling process. After t_r trajectory falls into classically allowed region in real time axis. So we can replace the action in Eq. (2.8) by

$$W^{re} = -\int_{t_r}^{\infty}\left[\frac{1}{2}\mathbf{v}^2(t) - \frac{Z}{r(t)} + I_p\right]dt, \tag{2.14}$$

$$W^{sub} = -\int_{t_s}^{t_r}\left[\frac{1}{2}\mathbf{v}^2(t) + I_p\right]dt, \tag{2.15}$$

$$W = W^{re} + W^{sub} \tag{2.16}$$

And the motion after t_r is governed by

$$\frac{d\mathbf{v}}{dt} = -\mathbf{E}(t) - \frac{Z\mathbf{r}}{r^3}. \tag{2.17}$$

Now the problem remained is how to deal with the initial condition for the classical trajectory in classically allowed region. It was suggested [50] that

$$\mathbf{v}(t_r) = \mathbf{p} + \mathbf{A}(t_r) \tag{2.18}$$

and

$$\mathbf{r}(t_r) = Re(\int_{t_s}^{t_r}(\mathbf{p} + \mathbf{A}(t))dt) = Re(\int_{t_s}^{t_r}\mathbf{A}(t)dt). \tag{2.19}$$

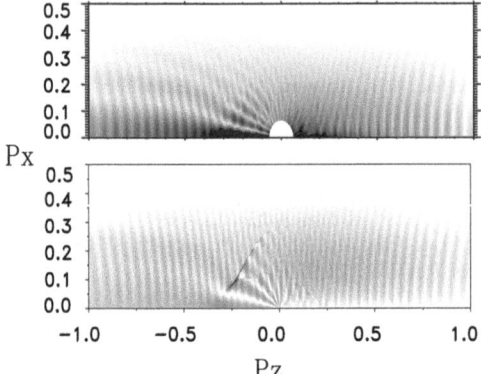

Fig. 2.8 Momentum distribution for photoelectrons from our classical trajectory ensemble simulation (*lower*) and TDSE result (*upper* Reprinted from[50],Copyright 2010,with permission from American Physical Society), respectively. The Z is polarization direction, X is lateral direction. The calculation is made for hydrogen atom irradiated by ultrashort laser pulse of intensity $100\,\text{TW/cm}^2$ and wavelength $2\,\mu m$. The "hole" structure at origin in quantum calculation arises from numerical truncation

The key point in the above scheme is to assign a proper quantum phase to each classical trajectory, so that interference effect can be include in the classical trajectory description. However, the above scheme does not account for the Coulomb effect in sub-barrier regime. It can be improved as follows. We randomly set a pair of tunneling time t_r and the lateral momentum $p_\perp$. Using the saddle point Eq. (2.13), we can obtain the imaginary time $t_i = Im(t_s)$ and the longitude momentum $p_\parallel$. The initial velocity after tunneling can be still determined by Eq. (2.18), but for the tunneling position we use the result from the semiclassical calculation in parabolic coordinates (see Sect. 2.1). After laser is over, the final momentum of the orbit is denoted as $\tilde{\mathbf{p}}$. Furthermore, we weigh each trajectory as $w(t_r, p_\perp)$ according to ADK theory presented in Eq. (2.3). Then, the transition amplitude from the ground state to the continuum state of asymptotic momentum $\tilde{\mathbf{p}}$ is given by

$$M_{\tilde{\mathbf{p}}} \sim \sum_{t_r, p_\perp} \sqrt{w(t_r, p_\perp)} \exp\left(i\,\mathrm{Re}(W(t_s(t_r, p_\perp)))\right), \qquad (2.20)$$

where $t_s(t_r, p_\perp)$ can be solved from the saddle point Eq. (2.13).

As illustration of the classical trajectory interference, we have calculated the momentum distribution of the photoelectrons of hydrogen atom irradiated by intense laser beam, as shown in Fig. 2.8. Our classical trajectory simulation reproduces the interference fringe structure and agrees to the quantum results from directly solving TDQE.

References

1. Ammosov, M.V., Delone, N.B., Krainov, V.P.: Zh. Eksp. Teor. Fiz. **91**, 2008 (1986)
2. Delone, N.B., Krainov, V.P.: J. Opt. Soc .Am. B. **8**, 1207 (1991).
3. Perelomov, A.M., Popov, V.S., Terentev, M.V.: Sov. Phys. JETP. **23**, 924 (1966)
4. Corkum, P.B.: Phys. Rev. Lett. **71**, 1994 (1993)
5. Landau, L.D., Lifshitz, E.M.: Quantum Mechanics, pp. 293. Pergamon Press (1977)
6. Hu, B., Liu, J., Chen, S.G.: Phys. Lett. A **236**, 533 (1997)
7. Paulus, G.G., et al.: Phys. Rev. Lett. **72**, 2851 (1994)
8. Yang, B., et al.: Phys. Rev. Lett. **71**, 3770 (1993)
9. DiMauro, L.F., Agostini, P.: Adv. Atomic Molecular Opt. Phys. **35**, 79 (1995) [4]
10. van Linden, H.B., van den Heuvell, Muller, H. G.: Multiphoton Processes. In: Smith, S.J., Knight, P. L. (eds.). Cambridge University Press, Cambridge (1988)
11. Gallagher, T. F.: Phys. Rev. Lett. **61**, 2304 (1988)
12. Corkum, P. B., Burnett, N. H., Brunel, F.: Phys. Rev. Lett. **62**, 1259 (1989)
13. Becker, W., Lohr, A., Kleber, M.: J. Phys.B **27**, L325 (1994)
14. Bao, D., Chen, S.-G., Liu, J.: Appl. Phys. B. **62**, 313 (1996)
15. Walker, B., et al.: Phys. Rev. Lett. **77**, 5031 (1996)
16. Paulus, G.G., Nicklich, W., Walther, H.: Europhys. Lett. **27**, 267 (1994)
17. Paulus, G.G., et al.: J. Phys.B **27**, L703 (1994)
18. Liu, J., Chen, S-G., Hu, B.: Acta Sinica Physica **7**, 89 (1998)
19. Liu, J., Chen, S. G., Li, B., Hu, B.: Chin. Phys. **9**, 24 (2000)
20. Gavrila, M.: J. Phys. B **35**, R147 (2002)
21. Su, Q., Eberly, J. H.: J. Opt. Soc. Am. B **7**, 564 (1990)
22. Eberly, J. H., Kulander, K. C.: Sci. **262**, 1229 (1993)
23. Dörr, M., Potvliege, R.M., Shakeshaft, R.: Phys. Rev. Lett. **64**, 2003 (1990)
24. Kulander, K.C., Schafer, K.J., Krause, J.L.: Phys. Rev. Lett. **66**, 2601 (1991)
25. de Boer, M. P., et al.: Phys. Rev. Lett. **71**, 3263 (1993)
26. Fedorov, M. V., Movsesian, A. M.:The traditional atomic stabilization requires that the laser frequency is higher than or at least comparable with the bound energy, which is usually not satisfied for the atoms in ground state. The stabilization issue is extended to the Rydberg atoms. J. Phys. B 21, L155 (1998)
27. Piraux, B., Potvliege, R. M.: Phys. Rev. A **57**, 5009 (1998)
28. Pont, M., Shakeshaft, R.: Phys. Rev. A **44**, R4110 (1991)
29. Askeland, S., et al.: Phys. Rev. A **84**, 033423 (2011)
30. Gavrila, M., Simbotin, I., Stroe, M.: Phys. Rev. A **78**, 033404 (2008)
31. Stroe, M., Simbotin, I., Gavrila, M.: Phys. Rev. A **78**, 033405 (2008)
32. Keldysh, L. V., Eksp, Zh.: Teor. Fiz. **47, 1945 (1964) [Sov. Phys. JETP 20, 1307 (1965)].**
33. Kulyagin, R.V., Taranukhin, V.D.: Laser Phys. **3**, 644 (1993)
34. Gavrilenk, V.P., Oks, E.: Can. J. Phys. **89**, 849 (2011)
35. Liu, H., et al.: PRL **109**, 093001 (2012)
36. Nubbemeyer, T., et al.: Phys. Rev. Lett. **101**, 233001 (2008)
37. Doron, E., Smilansky, U., Frenkel, A.: Quantum Chaotic Scattering and Microwave Experiments in Quantum Chaos, Enrico Fermi (1996), G. Casati, I. Guarneri, and U. Smilansky, (eds.) North-Holland (1993).
38. Eichmann, U., Nubbemeyer, T., Rottke, H., Sandner, W.: Nature **461**, 1261 (2009)
39. Maher-McWilliams, C., Douglas, P., Barker, P.F.: Nat. Photon. **6**, 386 (2012)
40. Qiu, M., et al.: Science **311**, 1440 (2006)
41. Gilijamse, J.J., Hoekstra, S., van de Meerakker, S.Y.T., Groenenboom, G.C., Meijer, G.: Sci. **313**, 1617 (2006)
42. Scharfenberg, L., van de Meerakker, S.Y.T., Meijer, G.: Phys. Chem. Chem. Phys. **13**, 8448 (2011)
43. Krems, R.V.: Phys. Chem. Chem. Phys. **10**, 4079 (2008)

44. Johnson, K.S., et al.: Sci. **280**, 1583 (1998)
45. Meschede, D.: J. Phys. Conf. Ser. **19**, 118 (2005)
46. Anderson, B.P., Kasevich, M.A.: Sci. **282**, 1686 (1998)
47. Landsman, A. S., Pfeiffer, A. N., Smolarski, M., Cirelli, C., Keller, U.: arXiv:1111.6036
48. Huang, K-y., Xia, Q-z.: Phys. Rev. A **87**, 033415 (2013)
49. Nubbemeyer, T., Gorling, K., Saenz, A., Eichmann, U., Sandner, W.: Phys. Rev. Lett. **101**, 233001 (2008)
50. Yan, T.-M., Popruzhenko, S.V., Vrakking, M. J. J., Bauer, D.: Phys. Rev. Lett. **105**, 253002 (2010)
51. Milošević, D. B., Paulus, G. G., Bauer, D., Becker, W.: J. Phys. B **39**, R203 (2006).
52. Popov, V.S.: Phys. Uspekhi **47**, 855 (2004)

Chapter 3
Double Ionization in Strong Laser Fields

Abstract Double ionization (NSDI) of atoms subject to ultrashort intense laser is a prototype model for the study of the three-body Coulomb problem intervened by the highly nonlinear interaction of electrons with a strong laser field. In this Chapter, we present some of its interesting aspects such as "knee" structure of double ionization yields, recollision threshold as well as some striking correlated electron momentum distributions, and provide insight into them with the help of semiclassical simulations based on classical trajectory ensemble.

3.1 Introduction: "Knee" Structure of Double Ionization Yields and Correlated Electron Momentum Distribution

Since the excessive double ionization was observed in helium experiments [1–3], much attention has been drown to multi-electron dynamics in laser-atom interactions. In these experiments the single ionization yield of He in a linearly polarized field is accurately predicted by the Single active electron (SAE) approximation. However, the case of double ionization is more complicated. In the regime of very high intensities [e.g., $I > 4 \times 10^{15}$ W/cm^2 in Fig. 3.1a], where strong double ionization occurs, the measured doubly charge ion yield are in good agreement with sequential SAE models. The experimental data, however, deviates seriously from the sequential SAE model and shows great enhancement in the knee regime $(0.8 - 3.0) \times 10^{15}$ W/cm^2, where the He^{2+}/He^{1+} yield ratio is close to a constant: 0.002, as indicated by the knee structure in Fig. 3.1a. This surprisingly large yield of double ionization obviously indicates that sequential ionization is no longer the dominant process in this regime and that electron-electron correlation has to be taken into account. Until recently it has been consensus that rescattering is the dominant mechanism for Nonsequential double ionization (NSDI) [4]. The electron recollision picture as a cornerstone of the rescattering mechanism inspires the further investigations that achieve insight into the microscopic dynamics of the ionization process on the timescale of

J. Liu, *Classical Trajectory Perspective of Atomic Ionization in Strong Laser Fields*, SpringerBriefs in Physics, DOI: 10.1007/978-3-642-40549-5_3, © The Author(s) 2014

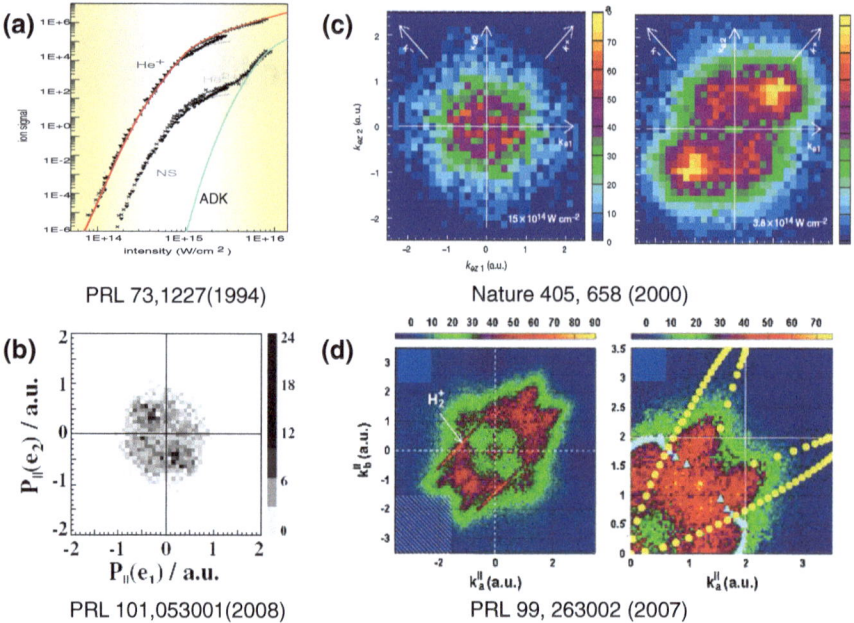

PRL 73,1227(1994)   Nature 405, 658 (2000)

PRL 101,053001(2008)   PRL 99, 263002 (2007)

Fig. 3.1 **a** Dots and crosses show measured ion yields of He^+ and He^{2+}. Red and green solid lines represent the results of ADK theory. The experimental results display the the typical "knee" structure of NSDI (reprinted from [2], Copyright 1994, with permission from American Physical Society). The other panels show some typical correlated electron momentum distribution along the laser polarization for NSDI of (**b**), argon at laser intensity of 4.0×10^{13} W/cm^2(cited from [5], Copyright 2008, with permission from American Physical Society) (**c**), argon at 15×10^{14} W/cm^2 and 3.8×10^{14} W/cm^2, respectively (reprinted from [6], Copyright 2000,with permission from Nature Publishing Group), and (**d**) helium at 4.5×10^{14} W/cm^2(cited from [7], Copyright 2007, with permission from American Physical Society). See text for details

subfemtosecond. Despite the great success of this picture, a comprehensive understanding of the microscopic dynamics in this recollision process is far from being complete.

On the other aspect, the advances in experimental techniques represented by the sophisticated Cold target recoil ion momentum spectroscopy (COLTRIMS) and high repetition-rate lasers, make it possible to record correlated electron momentum distributions with unprecedented high resolution. More and more interesting experimental phenomena emerge. For instance, a striking finger-like (or V-shaped) structure is observed for helium at laser intensity of $I = 4.5 \times 10^{14}$ W/cm^2 [Fig. 3.1d][7]. More interestingly, well below the recollision threshold (i.e., return electron energy is not enough to directly kick out the inner electron), strongly correlated back-to-back emission of the electrons along the polarization direction is observed that is in striking contrast to all previous data obtained above the recollision threshold [Fig. 3.1b][5]. These patterns in the momentum distribution plane can be regarded as "footprint" of the Coulomb three-body collision intervened by the non-perturbative laser field. It provides unprecedented access to the inner workings of atoms and calls for thorough investigation from theoretical sides.

3.2 Semiclassical Model for Nonsequential Double Ionization

Most experiments on double ionization are mainly confined to the tunneling regime, i.e., the ratio between the tunneling time of the outer electron and the inverse optical frequency (Keldysh parameter) is less than 1. With this consideration, we extend the previous semiclassical model for single ionization to double ionization case [8], see Fig. 3.2. The outer electron's tunneling ionization is discussed within quasistatic model, where the reduced 1D effective Schrödinger equation takes the following form [9],

$$\frac{d^2\phi}{d\eta^2} + \left[\frac{I_{p1}}{2} + \frac{1}{2\eta} + \frac{1}{4\eta^2} + \frac{1}{4}E(t_0)\eta\right]\phi = 0. \tag{3.1}$$

Fig. 3.2 The dashed and dotted lines correspond to the single ionization yields of He and He$^+$ predicted by ADK tunneling ionization; the full circles correspond to the results from our model. Inset: Intensity dependence of He^{2+}/He$^+$ ratio given by our model. The solid line is from the experiment [2]. Reprinted with the permission from Ref [8]. Copyright 2001, American Physical Society

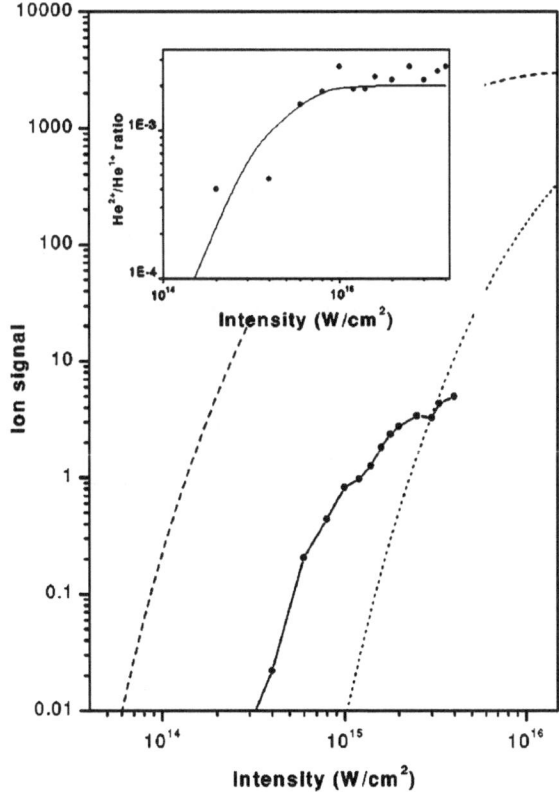

in which $E(t_0)$ is the instantaneous electric field. I_{p1} is the first ionization of atom. The above equation has the form of a one-dimensional Schrödinger equation with the potential $U_{\text{eff}} = -\frac{1}{4\eta} - \frac{1}{8\eta^2} - \frac{E(t_0)\eta}{8}$ and the effective ionization energy becomes $K = -\frac{I_{p1}}{4}$. The turning point η_0 at the outer edge of the distorted Coulomb potential is determined by $U_{\text{eff}}(\eta_0) = K$, which represents the location where the tunneling electron is born. Suppose the instantaneous field is along z axis, the initial coordinates of the tunneling electron will be $x_0 = y_0 = 0$, $z_0 = -\eta_0/2$. The tunneled electron has a Gaussian distribution on transverse velocity and zero longitudinal velocity, i.e., $v_z(t_0) = 0$, $v_x(t_0) = v_{x0}$, $v_y(t_0) = v_{y0}$.

The spherically symmetric ground-state He^+ is represented by microcanonical distribution, that was used in the Classical-trajectory monte carlo (CTMC) method [10–12]:

$$\rho(\mathbf{r}_2, \mathbf{p}_2) = \frac{\delta[I_{p2} - H_0(\mathbf{r}_2, \mathbf{p}_2)]}{K}. \tag{3.2}$$

Where $H_0(\mathbf{r}_2, \mathbf{p}_2) = p_2^2/2m_e - Ze^2/r_2$, I_{p2} the second ionization energy of atom and K is the normalization constant. This state is specified by the binding energy of the electron in the target atom and five additional parameters randomly distributed in the following ranges: ε is the eccentricity of the orbit with ε^2 randomly distributed in $[0, 1]$. u is the eccentric angle which has a complex distribution. In general, another new geometrical parameter is induced $\theta_n = u - \varepsilon \cos u$. It can be proved that θ_n is proportional to time, thus $0 \ll \theta_n \ll 2\pi$ corresponds to a periodic motion of an electron. If ε and θ_n are set, u can be obtained numerically. Finally, we get the initial coordinates and momenta of the electron. ϕ, θ, η are Euler angles, ($-\pi \ll \phi \ll \pi$, $-1 \ll \cos\theta \ll 1$, $-\pi \ll \eta \ll \pi$). A random distribution of these parameters corresponds to equal probability of the inner electron having any phase in its periodic motion.

The evolution of the two electrons after the first electron has tunneled and the electron-electron interaction are described by the classical equations,

$$\frac{d^2\mathbf{r}_1}{dt^2} = -\frac{2\mathbf{r}_1}{r_1^3} + \frac{\mathbf{r}_1 - \mathbf{r}_2}{|\mathbf{r}_1 - \mathbf{r}_2|^3} - \mathbf{E}(t), \tag{3.3}$$

$$\frac{d^2\mathbf{r}_2}{dt^2} = -\frac{2\mathbf{r}_2}{r_2^3} - \frac{\mathbf{r}_1 - \mathbf{r}_2}{|\mathbf{r}_1 - \mathbf{r}_2|^3} - \mathbf{E}(t). \tag{3.4}$$

After the laser pulse, double ionization events are identified according to energy criterion and collected for statistics. Each event is weighed by the tunneling rate of the outer electron at instantaneous field, i.e., $w(t_0, v_\perp) = w(0)w(1)$, with

$$w(0) = \left[\frac{E(t_0)}{4}\right]\left[\frac{4\kappa^4}{E(t_0)}\right]^{2/\kappa} e^{-\frac{2\kappa^3}{3E(t_0)}}, \quad w(1) = \frac{\kappa}{\pi E(t_0)} e^{-\kappa(v_{x0}^2 + v_{y0}^2)/E(t_0)}, \tag{3.5}$$

in which the parameter $\kappa = \sqrt{2I_{p2}}$. Concerned physical quantities can be obtained through averaging on trajectory ensemble.

The above semiclassical model has been applied to helium. Without any fitting parameters we can reproduce the observed "knee" structure in He^{2+}/He^{1+} yield ratio and the saturated ratio yield of 0.002 quantitatively [8]. In the following, we mainly address the intriguing patterns appeared in correlated electron momentum distribution using the developed semiclassical model.

3.3 Finger-Like Structure in the Correlated Electron Momentum Distribution

New high resolution and high statistics COLTRIMS experiments on Double ionization (DI) of helium are performed independently by two groups and a striking finger-like (or V-shaped) structure is observed [13, 14] in the correlated electron momenta parallel to the laser polarization. The observation is in qualitative accordance with the prediction of S-matrix approach [15, 16] and quantum mechanical calculation [17, 18]. However, in the S-matrix model, a rather unphysical contact-type electron-electron interaction yields the agreement with the experimental data, while, the implementation of a realistic Coulomb interaction results in clear deviations from the data. On the other aspect, the dynamical details of the electron recollision process is hardly extracted from the solution of the time-dependent Schrödinger equation. In this section, we address the influence of several physical mechanisms, such as the electron-electron and electron-ion interaction, on the finger-like structure, achieve insight into underlying subcyle trajectory dynamics by exploiting the developed semiclassical model.

3.3.1 Finger-Like Structure

The resulting electron momentum distribution, calculated with our semiclassical model for the same parameters as in the experiment [13], is shown in Fig. 3.3a. The calculation reproduces many key features observed in the experiment, including the emission of the two electrons primarily into the same hemisphere, the small circular accumulation around the zero momentum surrounded by four elliptical hard-to-reach regime, and more importantly, the finger-like structure beyond $2\sqrt{U_p}$ ($U_p = \varepsilon_0^2/4\omega^2$, is the ponderomotive energy). With assuming that the colliding electrons leave the atom with no significant energy and the motion of the individual electron is solely determined by the laser field, the parallel momentum $k_{1,2}^{||}$ of each electron results exclusively from the acceleration in the optical field. Within this simple model $2\sqrt{U_p}$ should be the maximum momentum and the momentum distribution favors accumulation in the diagonal zone because the electrons are emitted nearly simultaneously. However, it should be noted that $2\sqrt{U_p}$ is not the classical limit of the parallel momentum as clearly revealed by Fig. 3.3a and discussed in Ref. [19].

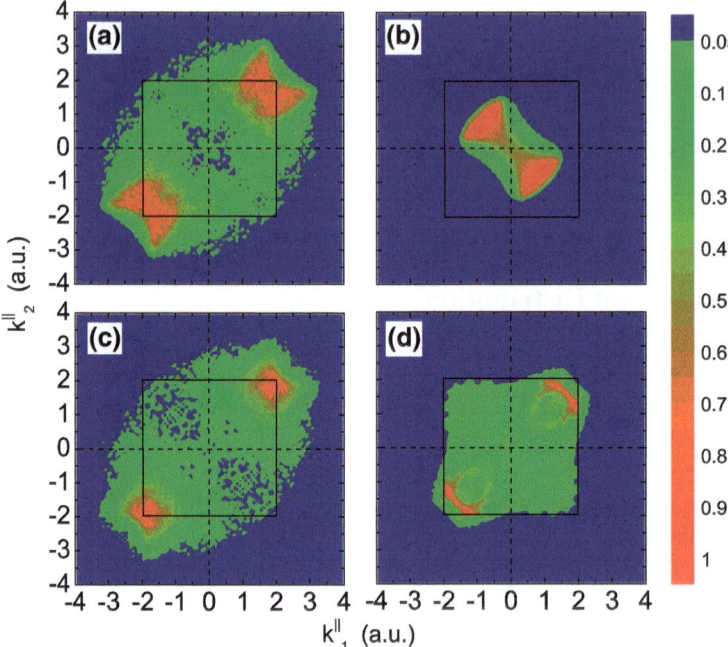

Fig. 3.3 **a** Distribution of correlated electron momenta along the laser polarization for Helium DI irradiated by 800 nm, 4.5×10^{14} W/cm^2 laser pulses. The black box indicates the $2\sqrt{U_p}$ boundary of electron momentum. The model calculations under various circumstances yield very different momentum distribution patterns (see text for details): **b** the laser field is removed and the tunneled electrons are replaced by a beam of projectile electrons; **c** the electron-electron Coulombic interaction is replaced with a Yukawa potential; **d** the nuclear Coulomb potential is softened. Reprinted with the permission from Ref [20]. Copyright 2008, American Physical Society

We now proceed to explore the physical effects that give rise to this peculiar finger-like structure, using the Classical-trajectory (CT) method. In the context of strong field double ionization, there are essentially three major effects that may play significant role in the double electron emission dynamics: electron-laser field interaction that occurs throughout the DI process, electron-nuclear Coulomb interaction in the post-collision duration and the inter-electron Coulomb repulsion which becomes significant when both electrons get close. Below we investigate all three interactions and clarify their distinct roles in the formation of the finger-like structure.

The first step is to check the role of the external laser field and an additional calculation is thus performed, in which the laser field is intentionally removed and the tunneled electrons are replaced by a beam of projectile ones with incident energy of $3.17U_p$, corresponding to the maximal kinetic energy of the tunneled electrons upon recollision. The result is shown in Fig. 3.3b. Two significant differences from the complete model calculation in Fig. 3.3a are found: (i) the finger-like structure beyond $2\sqrt{U_p}$ completely disappears; (ii) the two emitted electrons tend to distribute in the second and fourth quadrants, indicating that the incident electron transfers much of its momentum to the bound one while itself is back-scattered into the opposite direction. The comparison between Figs. 3.3a and b shows the most important role of the laser field in turning the two back-to-back emitted electrons into the same direction and the post-collision acceleration of the electrons resulting in the finger-like structure beyond $2\sqrt{U_p}$.

The next step concerns the question if this finger-like structure is a fingerprint of a strong inter-electron correlation between the ionizing electrons. We have performed another calculation in which the final-state electron Coulomb repulsion has been deliberately neglected by replacing the electron Coulombic interaction $V_{ee} = \frac{1}{|\mathbf{r}_{1,2}|}$ with Yukawa repulsion potential of the form $V_{ee} = \exp[-\lambda r_b]/r_b$, where $r_b = \sqrt{|\mathbf{r}_{1,2}|^2 + b^2}$, $\lambda = 5.0$ and $b = 0.2$. The result of this calculation [Fig. 3.3c] shows that the prominent finger-like structure is to a large extent reduced and thus provides a clear evidence that the final-state electron correlation plays a significant role.

Last but not the least procedure is to justify the role of the electron-nuclear interaction, which is commonly believed to be the main reason for the recoil collision in field-free (e, 2e) process. This interaction was also suggested to be the very ingredient for the field-assisted recoil collision in the context of intense field DI of atomic helium [13]. Accordingly, an additional calculation, in which we soften the nuclear Coulomb attraction by employing $V_{ne}^i = -2/\sqrt{|\mathbf{r}_i|^2 + a^2}$, where a is chosen as 1.0 to match the ground state energy of He$^+$, is performed.

Physically, the shielding of nuclear potential would to a great extent diminish the Coulomb focusing effect that have significant effects upon both electrons. Clearly, a Coulombic potential would attract the tunneled electron more dramatically when it moves near the atomic core. Such strong attraction may unambiguously bring the tunneled electron to share more kinetic energy with the bound one. For the bound electron, after achieving considerable transferred momentum upon collision, it may elastically backscatter from the Coulombic core on its way out of the atom. This double scattering process is coined as recoil collision [21] and was routinely found in traditional electron impact ionization experiments, especially when the projectile electron possesses the energy of only a few times of the binding energy of the inner one. The result shown in Fig. 3.3d indicates that Coulomb focusing effect is decisive for the production of the electrons with high energy, and thus the finger-like structure.

3.3.2 Trajectory Dynamics

It has been recognized that the characteristics of the DI trajectory can be well represented by the recollision and DI time [22–24]. We thus provide such an information for the trajectories that contribute to the finger-like structure in Figs. 3.4c and d. It is found that, in Fig. 3.4c, the electron pairs contributing to the finger-like structure tend to encounter right at zero field. Within rescattering picture, this can be understood as that these trajectories include the most energetic collisions for which the tunneled electrons are released at the laser phase of about 17°, and return around the zero crossing of the electric field at 270° with maximal energy of $3.17U_p$ [25, 26].

Upon recollision, the bound electron may be directly freed, a process termed as Collision ionization (CI), or be excited and subsequently ionized by the next field maximum, known as Collision-excitation ionization (CEI) [27]. The smaller and larger peaks in Fig. 3.4d correspond to these two mechanisms, respectively. The smaller ones around zero field represent the electron pairs emitted after a very short thermalization process ($\sim$attosecond) [28], and the larger ones correspond to CEI events with small time delay of 0.15T (i.e., 0.3π phase difference from the collision phase peaks) [29]. The results shown in Figs. 3.4c and d indicate that both CI and CEI with small time delay could contribute to the finger-like structure.

To further unveil the microscopic mechanism underlying this unusual pattern, we collect all trajectories that constitute the finger-like structure and trace back their dynamical evolution at collision and post-collision. With the CT diagnosis we sketch two types of trajectory configurations that are responsible for the asymmetric electron energy sharing after electron-electron collision and lead to the finger-like structure beyond $2\sqrt{U_p}$ (see Figs. 3.4a and b, correspond to type-I and II configurations, respectively). While the tunneled electron is driven back to the nucleus by the laser

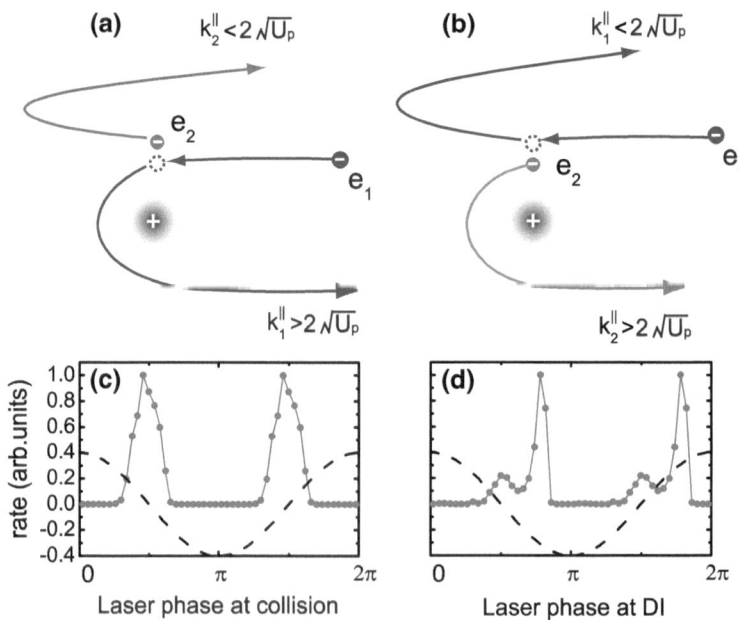

Fig. 3.4 a–b Two trajectory configurations responsible for the finger like structure. DI yield versus laser phase at recollision (**c**) and at DI moment (**d**). Here, the statistics is only collected for the DI trajectories that contribute to the prominent finger-like structure, i.e., the regimes of $1.5\,\text{a.u.} < |k_i^{||}| < 2.0\,\text{a.u.}$ and $2.5\,\text{a.u.} < |k_j^{||}| < 3.0\,\text{a.u.}$ where $i, j = 1, 2, i \neq j$. The dashed curves represent laser fields.Reprinted with the permission from Ref [20]. Copyright 2008, American Physical Society

field, the field strength reduces to zero and the collision is essentially a field free three-body system under the pure Coulomb potential. For type-I configuration of Fig. 3.4a, the electron-electron collision near nucleus could lead to the following consequences: the second electron acquires considerable momentum from the returned electron and emits in the forward direction, while the returned electron is slowed down. Under the influence of the nuclear attraction, the latter is transferred to a hyperbolic orbit around the nucleus with a large scattering angle. In this way, the returned electron reverses its direction in a time scale of attoseconds after the collision. Meanwhile the laser field changes its direction. As the returned electron has nonzero residual momentum parallel to the instantaneous field direction and is further accelerated by the field, its final longitudinal momentum is expected to be above $2\sqrt{U_p}$. For the second electron, after collision its initial momentum is opposite to the instantaneous field direction, one expects that its final longitudinal momentum is below $2\sqrt{U_p}$. In Fig. 3.4 b where a type-II trajectory is schematically shown, the situation is similar except that the roles of the two electrons have exchanged: Under assistance of nuclear Coulomb attraction, the second electron acquires a nonzero momentum parallel to the instantaneous field and therefore emits with a longitudinal momentum larger than $2\sqrt{U_p}$. Although the returned electron is slowed down by electron-electron repulsion, it still has a residual momentum opposite to the instantaneous field, resulting in a final longitudinal momentum below $2\sqrt{U_p}$. Our statistics reveals that the DI events in finger-like structure consists of 70 % type-I and 30 % type-II configuration. It indicates that for the trajectories contributing to the finger-like structure the slower electron is usually the one ejected from the ion ground state, in agreement with the quantum calculation [17].

3.4 Transition to Below the Recollision Threshold

3.4.1 Recollision Induced Excitation-Tunneling (RIET) Effect

In the rescattering picture, one electron is first released through quantum tunneling, then is driven back to its parent ion and imparts its kinetic energy to dislodge a second electron. The maximal energy of the returned electron is $3.17U_p$. When the idea of recollision was first introduced, it was anticipated that the double ionization yield would undergo a sudden drop suppose the maximum returned energy is smaller than the ionization potential of the inner electron, i.e., in the regime Below the recollision threshold (BRT). But, by now it is widely recognized that recollision excitation and field suppression effects might extend the process to lower intensities. More interestingly, a recent experiment in the deep BRT regime [5] has revealed that the two photoelectrons prefer to drift out in opposite directions with showing the so-called anticorrelated phenomenon, in contrast to all previous observations near or above the recollision threshold (see, for example, Ref. [30–32]). A comprehensive understanding of the physical origin of the above striking phenomena is far from complete, although some progress has been made towards the BRT regime using both purely classical approach [33] and quantum treatment [34]. In the section, we investigate the strong-filed double ionization at the transition to the BRT regime using an extended semiclassical model, in which the Recollision-induced excitation-tunneling (RIET) effect has been taken into account with the Wentzel-Kramers-Brillouin (WKB) approach.

The model introduced in Sect. 3.2 has been proved to work well in the regime above the recollision threshold [35, 36]. To extend it to the BRT regime, we need include the RIET effect in the second step [37]. This is done by allowing the bound electron to tunnel through the potential barrier whenever it reaches the outer turning point, where $p_{i,z} = 0$ and $z_i \varepsilon(t) < 0$, with a tunneling probability P_i^{tul} given by the WKB approximation

$$P_i^{tul} = \exp\left[-2\sqrt{2} \int_{z_i^{in}}^{z_i^{out}} \sqrt{V(z_i) - V(z_i^n)} dz_i\right]. \tag{3.6}$$

Here, z_i^{in} and z_i^{out} are the two roots ($\left|z_i^{out}\right| > \left|z_i^{in}\right|$) of the equation for z_i, $V(z_i) = -2/r_i + z_i \varepsilon(t) = -2/r_i^{in} + z_i^{in} \varepsilon(t)$ [38, 39].

It is well-known that when the tunneled electron is thrown back to the parent core, its maximal kinetic energy is $3.17U_p$. As the laser intensity gradually decreases, the maximal recolliding energy becomes insufficient to directly ionize the inner electron. This is the case for the lowest laser intensity in the above calculations, where $3.17U_p = 0.28$ a.u. is far below $|I_{p2}| = 1.02$ a.u.. As one may expect, the inner electron that has been excited by the first recollision requires other mechanisms to liberate it, for example, by means of a second (even multiple) recollision or the field-assisted tunneling ionization. The former can be straightforward modeled classically [33]. We term it as the Recollision-induced direct ionization (RIDI). The latter, i.e., the (RIET), is of quantum nature and out of the capability of previous classical treatment [33]. These two different mechanisms leave distinct footprints in the correlated momentum spectra. This can be clearly seen by comparing the second and third columns of Fig. 3.5, where we split the total correlated momentum spectra into two parts according to different emission mechanisms.

From the mid-column of Fig. 3.5 for the RIDI mechanism, we see that the correlated momentum spectra exhibit a clear transition from the dominance of correlated emission to a situation where anti-correlated ejection has become much stronger and even the most important contribution, when the laser intensity decreases to the BRT regime. The patterns are to some extent analogous to that in the first column for the total distribution.

The physical picture for the double ionization at the transition to the BRT regime is summarized in the percentile map of Fig. 3.6. Our results unveil that: (i) both the mechanisms of RIDI and RIET significantly contribute to the double ionization yield; (ii) the RIDI mechanism plays a decisive role for the transition from correlation to anticorrelation; (iii) the RIET mechanism always prefers to cause back-to-back

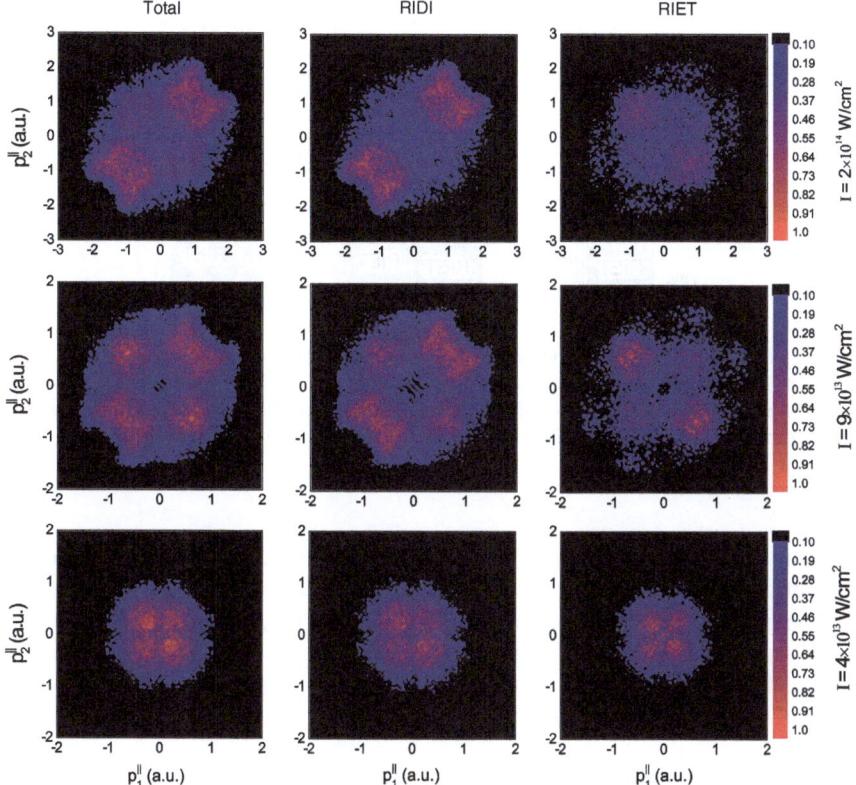

Fig. 3.5 Correlated momentum spectra for double ionization of argon at different laser intensities ranging from above to below the recollision threshold. Shown are the momentum components $p_{1,2}^{\parallel}$ along the laser polarization direction. The second and third columns are deduced from the first column by superimposing additional restriction on our statistics according to different emission mechanisms.Reprinted with the permission from Ref [37]. Copyright 2010, American Physical Society

emission. Its percentage over the total NSDI events increases monotonically as the laser intensity decreases.

3.4.2 Modified Recollision Threshold

After the recollision, the kinetic energy of the returned electron will be shared by the struck one, i.e., $E_r + E_s \simeq 3.17U_p - |I_{p2}|$. Here, E_r and E_s denote the energy of returned electron and struck electron after recollision, respectively. Thus, single recollision induced DI becomes possible only when both electrons can pass over the suppressed barrier, i.e., $\min(E_r/4, E_s/4) \geq V_b$. Obviously, the solution of the above

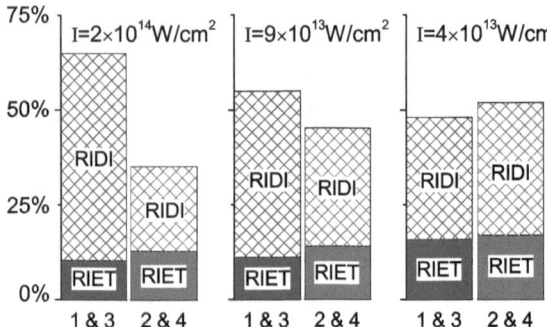

Fig. 3.6 Percentile map for doubly ionizing trajectories at three different laser intensities. All trajectories are classified into two categories based on whether the inner electron is freed through RIDI (*hatched area*) or RIET (*color-filled area*). The numbers at the bottom denote the quadrants of the correlated momentum plane.Reprinted with the permission from Ref [37]. Copyright 2010, American Physical Society

inequation depends on how the energy be shared by the electrons after recollision. We thus introduce the asymmetry parameter $\kappa = |(E_r - E_s)/(E_r + E_s)|$ to characterize the energy apportion. We consider the regime that $3.17U_p - |I_{p2}| < 0$ and the both electrons populate bound states after the recollision. $\kappa = 0$ corresponds to equal energy sharing while $\kappa = 1$ corresponds to most uneven energy apportion. For a given κ, the threshold intensity will be determined through following equation,

$$3.17U_p - |I_{p2}| \simeq -4/(1+\kappa)\sqrt{\varepsilon_0}, \quad 0 \leq \kappa \leq 1. \tag{3.7}$$

In practice, the energy is usually not equally allotted to the returned electron and struck electron after recollision. Numerically, we have made statistics on the distribution probability with respect to the asymmetry parameter κ, where the parameters are chosen close to the threshold. It shows that, the distribution function has a broad peak around zero with a long tail and decreases to zero at $\kappa = 1$. According to our calculation, the main profile of the distribution function is general while some detailed structures vary with the field parameters and the atomic structure.

In general, Eq. 3.7 defines a narrow transition zone whose upper boundary corresponding to $\kappa = 1$ and lower boundary corresponding to $\kappa = 0$ in the plotting of intensity vs wavelength as shown in Fig. 3.7. Above the zone, the DI events are mainly induced by single recollision and therefore the side-by-side emission dominates, whereas, below the zone, single-recollision-induced DI is forbidden and the anticorrelation becomes prominent. The transition from correlation to anticorrelation is expected to emerge in the transition zone.

Fig. 3.7 Threshold intensity versus laser wavelength. Comparisons are made between various theories and experimental data taken from a series of literature: Green and red indicate that the observed pattern is correlation and anticorrelation, respectively. Reprinted with the permission from Ref [37]. Copyright 2010, American Physical Society

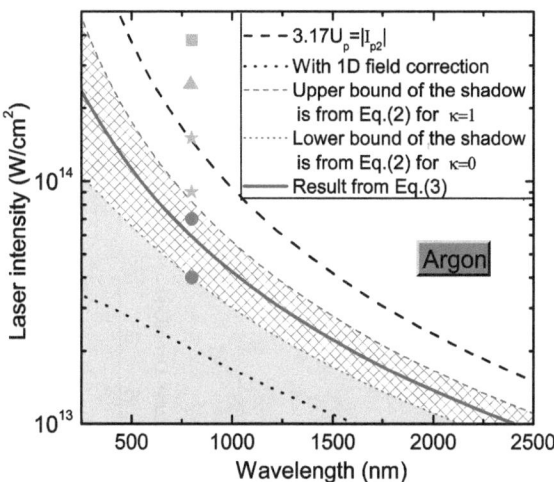

3.5 Double Ionization Dynamics of Diatomic Molecules

Compared with the atomic case, NSDI of molecules are even more complicated due to the additional molecular freedom[40, 41]. In addition to the typical strong-field phenomena like excessive DI yields in the tunneling regime, high harmonic generation of the driving laser field, and momentum correlation between the two emitted electrons, some new phenomena such as bond softening [42], zero photon dissociation [43], and alignment dependence of DI yield [44–46] were observed in recent molecular experiments. However, these experimental observations for molecular DI are far from well understood in theory. The complex dynamics of correlated $e - e$ pair responding to two-center nuclear attraction and laser force poses a great challenge to any quantum theoretical treatment. A fully-dimensional quantum-mechanical computation from first principle is very time consuming even for the simplest case of highly symmetric atoms [47]. This leaves approximate approaches developed recently, such as one-dimensional quantum model [48], many-body S-matrix [49] and simplified classical methods [50]. In either case, the complex electron dynamics which is crucial for molecular DI is not fully explored and the theoretical results can not account for experimental data quantitatively. In this section, we develop the feasible semiclassical theory to study the NSDI of molecules [29].

3.5.1 Enhanced Double Ionization Rate

Tunneling regime. In the long-wavelength limit, the laser field varies slowly in time and can be regarded as a quasi-static field compared with valence electron's circular motion around nuclei. Under this field, the Coulomb potential between nuclei

Fig. 3.8 a Tunneling ionization. **b** Over-the-barrier
ionization. **c** Contour plot of
the combined potential of the
nuclear Coulomb attraction
and the external laser field. It
clearly shows that the saddle
point locates approximately
along the direction of the
external field. Reprinted with
the permission from Ref
[52]. Copyright 2008,
American Physical Soci-
ety

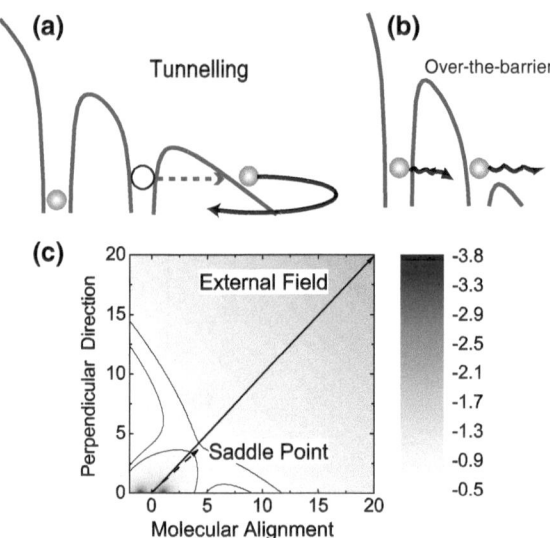

and electrons is significantly distorted. When the instantaneous field (at time t_0) is
sufficiently strong but still smaller than a threshold value (see Fig. 3.8a), one electron
is released at the outer edge of the suppressed Coulomb potential through quantum
tunneling with a rate $\varpi(t_0)$ given by molecular ADK formula[51].

The electron tunnels out through a saddle point [53] directing to a channel of the
local minimum in the combined potential of the nuclear Coulomb attraction and the
external laser field (see, Fig. 3.8c). Because the difference between the direction of
the saddle point and the external field is very small, we safely regard the external
field direction (z axis) as the tunneling direction. Thus, the initial position of the
tunneled electron can be derived from following equation,

$$-\frac{1}{r_{a1}} - \frac{1}{r_{b1}} + \int \frac{|\Psi(\mathbf{r}_2)|^2}{|\mathbf{r}_1 - \mathbf{r}_2|}d\mathbf{r}_2 + I_{p1} - z_1\varepsilon(t_0) = 0, \qquad (3.8)$$

with $x_1 = y_1 = 0$. The wavefunction Ψ is given by the Linear combination of
the atomic orbital-molecular orbital (LCAO-MO) approximation. Taking N_2^+ for
example, we choose $\phi(r) = \frac{\lambda^{3/2}}{\sqrt{\pi}}e^{-\lambda r}$ as the trial function to construct the molecular
orbital $\Psi(\mathbf{r}_2) = c[\phi(r_{a2}) + \phi(r_{b2})]$, where c is the normalization factor. The parameter λ, which equals to 1.54 for N_2^+, is determined through variational approach. That
is, calculate the variational energy for the given wavefunction and assume it equals
to the second ionization energy of the molecule. The initial velocity of the tunneled
electron is set to be $(v_\perp \cos\varphi, v_\perp \sin\varphi, 0)$, with $v_\perp$ having the same distribution as
that in atomic case, i.e.,

$$w(v_\perp)dv_\perp = \frac{2(2I_{p1})^{1/2}v_\perp}{\varepsilon(t_0)} \exp\left(-\frac{v_\perp^2(2I_{p1})^{1/2}}{\varepsilon(t_0)}\right)dv_\perp, \tag{3.9}$$

where φ is the polar angle of the transverse velocity uniformly distributed in the interval $[0, 2\pi]$.

In order to get the initial velocity distribution, we employ a technique widely used in Classical monte carlo (MC) simulation. Firstly, we generate two random number $v_\perp^{test}$ and w_{test} in the interval $[0, v_\perp^{max}]$ and $[0, w_{max}]$, respectively. If $w(v_\perp^{test}) > w_{test}$, $v_\perp^{test}$ is kept as the initial transverse velocity, otherwise it is rejected and the above procedure is repeated.

For the bound electron, the initial position and momentum are depicted by Single-electron microcanonical distribution (SMD) ,

$$F(\mathbf{r}_2, \mathbf{p}_2) = k\delta[I_{p2} - \mathbf{p}_2^2/2 - W(r_{a2}, r_{b2})], \tag{3.10}$$

where k is the normalization factor, I_{p2} denotes the ionization energy of molecular ions such as N_2^+, and $W(r_{a2}, r_{b2}) = -1/r_{a2} - 1/r_{b2}$ is the total interaction potential between the bound electron and two nuclei.

Over-the-barrier regime. To give a complete description of the DI of molecular system for the whole range of laser intensities (see Fig. 3.9), one needs to further extend the above model to over-the-barrier regime (Fig. 3.8b, in such case Eq. (3.8) has no real roots). This is done by constructing the initial conditions with double-electron microcanonical distribution (DMD) [54], i.e.,

$$F(\mathbf{r}_1, \mathbf{r}_2, \mathbf{p}_1, \mathbf{p}_2) = \frac{1}{2}[f_\alpha(\mathbf{r}_1, \mathbf{p}_1)f_\beta(\mathbf{r}_2, \mathbf{p}_2) + f_\beta(\mathbf{r}_1, \mathbf{p}_1)f_\alpha(\mathbf{r}_2, \mathbf{p}_2)], \tag{3.11}$$

with

$$f_{\alpha,\beta}(\mathbf{r}, \mathbf{p}) = k\delta[I_{p1} - \frac{\mathbf{p}^2}{2} - W(r_a, r_b) - V_{\alpha,\beta}(\mathbf{r})], \tag{3.12}$$

where $V_{\alpha,\beta}(\mathbf{r})$ represents the mean interaction between two electrons, $V_{\alpha,\beta}(\mathbf{r}) = \frac{1}{r_{b,a}}$ $[1 - (1 + \kappa r_{b,a})e^{-2\kappa r_{b,a}}]$, κ can be obtained by a variational calculation of the ionization energy of molecules ($\kappa = 1.14$ for N_2). Details can be found in Ref. [54].

We would like to give some remarks here. (i) In our calculations, the trajectories obtained from DMD are weighed by $\varpi(t_0)$ [51]. (ii) Part of the electrons obtained from DMD could "self-ionize" even without the presence of external field. To avoid such an unphysical self-ionization, we evolute the electrons for several optical cycles freely and abandon those samples whose energy are greater than zero during the free evolution process. (iii) The initial total energy distribution has a long tail on both sides. They are cut off by introducing two parameters E_{min} and E_{max} [55], which satisfy

$$\frac{\int_{E_{min}}^{E_{max}} E\rho(E)dE}{\int_{E_{min}}^{E_{max}} \rho(E)dE} = E_M, \tag{3.13}$$

Fig. 3.9 Comparison between DI data [41] and theory for nitrogen molecule [29]. $0.185\,PW/cm^2$ is the threshold intensity separates the tunneling regime and over-the-barrier regime as schematically plotted in Fig. 3.8. Reprinted with the permission from Ref [29]. Copyright 2007, American Physical Society

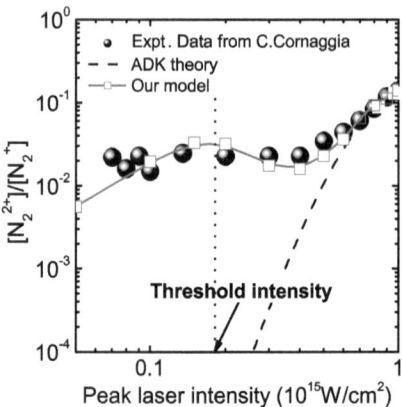

where $E_M \approx I_{p1} + I_{p2}$ is the most probable energy and $\rho(E)$ is the state density around E. In the following calculations, we choose $E_{max} - E_{min} = 0.2\,a.u.$.

With the above initial conditions, the Newtonian equations are solved using the 4-5th step-adaptive Runge-Kutta algorithm and DI events are identified by energy criterion. In our calculations, more than 10^5 weighted (i.e. by rate $\varpi(t_0)$) classical trajectories of electron pair are traced and a few thousands or more of DI events are collected for statistics. Convergency of the numerical results is further tested by increasing the number of launched trajectories twice.

In Fig. 3.9, the ratio between double and single ionization yield is plotted with respect to the peak laser intensities from 5.0×10^{13} W/cm^2 to 1.0×10^{15} W/cm^2. Our numerical results show good agreement with the experimental data over the whole range. The curve can be clearly divided into three parts: the low intensity regime (5.0×10^{13} W/cm^2 to 1.85×10^{14} W/cm^2), where tunneling ionization plays an important role; the very high intensity regime (5.0×10^{14} W/cm^2 to 1.0×10^{15} W/cm^2) which can be also well described by ADK theory, indicating that the electrons are pulled out sequentially. In particular, there exists a plateau (moderate intensity) regime , which can be seen from the corresponding experiments in Ref. [41]. Back analysis about classical trajectories indicates that the underlying dominant mechanisms in these three regimes are widely divergent as will be shown latter.

3.5.2 Typical Classical Trajectories

In the tunneling regime that the laser field significantly suppresses the Coulomb barrier but not strong enough to directly draw out the electron. In this case, the tunneled electron moves around the molecular core for some time, then it is driven by the oscillating laser field (Fig. 3.10a) to collide with the bound electron near the cores (Fig. 3.10b). As a result of the hard collision, a big change in the electrons'

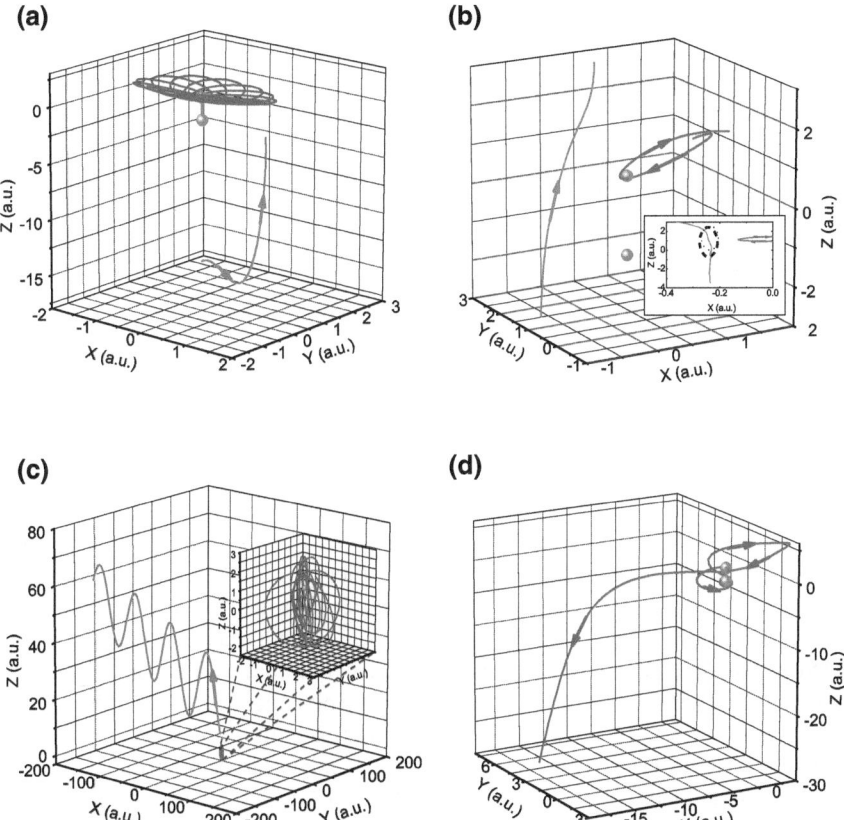

Fig. 3.10 (*Color online*). A typical CEI trajectory in the tunneling regime is divided into four stages: **a** *travel stage*, the tunneled electron travels freely in the laser field while the bound electron rotates around one of the two cores; **b** *recollision stage*, the two electrons come close and collide with each other, and then diverge with a large lateral recoiled momentum; **c** *excited stage*, after the collision the returned electron is driven away from the cores by the laser field while the bound electron still stays in the vicinity of the molecular cores. The increase in the radius of the classical orbits indicates the excitation of the bound electron; **d** *2nd ionization stage*, the excited electron will be ionized subsequently at the time when the laser field lowers the Coulomb barrier significantly and electron acquires enough energy from the laser field.Reprinted with the permission from Ref [52]. Copyright 2008, American Physical Society

orbits (highlighted by the black dash line) is witnessed by the inset of Fig. 3.10b. This intimate contact could lead to the instant ($\sim$attosecond) ionization of both two electrons (Fig. 3.11) or the excitation of the bound electron (Fig. 3.10c).

For Collision-ionization (CI), the incident electron knock out the inner electron while itself is bounced back with almost 180° inversion. Two electrons gain equal drift momentum and then emit in the same direction. The ionized electrons with this kind of trajectories exhibit a strong correlation and are responsible for the bright spots in the first and the third quadrants of the parellel momentum plane.

Fig. 3.11 (*Color online*).
Typical CI trajectory in the
tunneling regime. The incident
electron quickly knock out the
inner electron while itself
is back scattered at about
180°. Both electrons obtain
equal drifted momentum and
emit in the same direction
simultaneously.Reprinted
with the permission from
Ref [52]. Copyright 2008,
American Physical Sciety

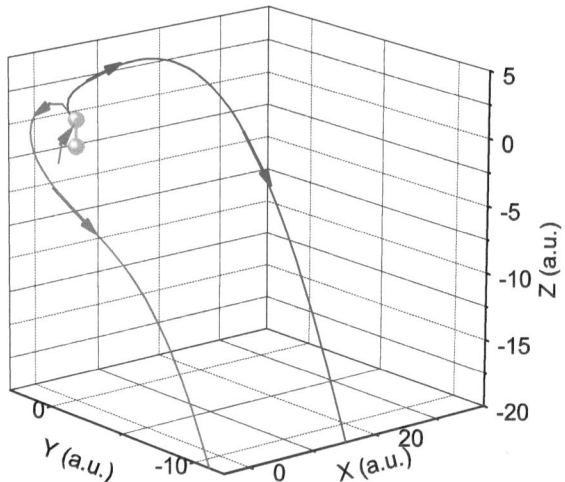

In the above CI type of the DI, the kinetic energy of the returned electron need to exceed the ionization energy of the bound electron and the electrons close to each other sharing the excess kinetic energy. On the other aspect, in the tunneled electron was simulated by a Gaussian wave packet of initial width of around 0.5 a.u., evolving and expanding in the combined Coulomb potential and the laser field. It was shown that, when the tunneled electron is driven back to the vicinity of the cores, its width turns to be several tens of a.u. that is much larger than the size of the inner electron cloud (about 1.0 a.u., see e.g. Fig. 3.10a). This means that, usually the returned electron cannot directly impact the bound electron but just "softly" collide with it, leading to the excitation rather than the "instant ionization" of the inner electron. In this case, the CEI type of DI emerges. For the CEI, the inner electron is firstly excited through the soft collision with the returned electron and then undergoes a time-delayed ($\sim$ a few optical periods) field-assisted ionization. After the collision, the returned electron escapes out and quivers in the laser field (red line in Fig. 3.10c) while the excited electron is still bounded around the cores (blue line in Fig. 3.10c). We notice that before the collision the bound electron rotates tightly around one core (blue line in Fig. 3.10a) while after collision the cloud of the electron motion expends to around two cores (blue line in Fig. 3.10c). The excited electron will be ionized subsequently at the time when the laser field lowers the Coulomb barrier significantly and it absorbs enough energy from the laser field. (Fig. 3.10d).

In the plateau regime where the peak laser intensity exceeds the threshold value of 0.185 PW/cm^2, over-the-barrier ionization occurs for certain instantaneous fields. In this case, we observe more complicated trajectories involved with multiple collisions. The difference between the trajectories in this regime and that in the tunneling regime is the emergence of an initially entangled stage. In this stage, the two electrons rotate around each of the two cores separately and experience multiple collisions. The collisions occurred in the region between the two cores are slight and only give rise

Fig. 3.12 (*Color online*). The typical trajectories in over-the-barrier regime. See text for detailed description. Reprinted with the permission from Ref [52]. Copyright 2008, American Physical Society

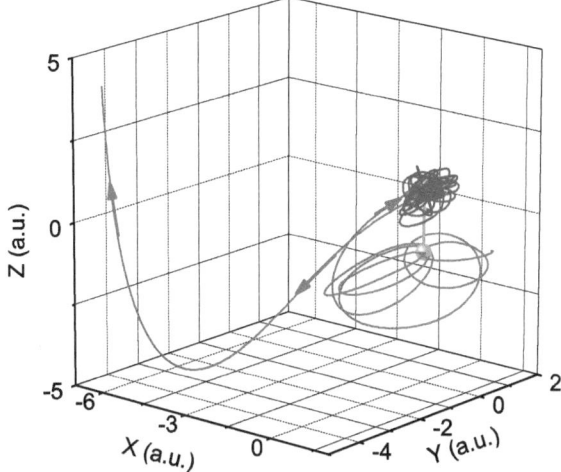

to small energy exchange between the two electrons. When one electron enters into the vicinity of the other core with the assistance of laser field, the collision between two electrons will be hard and then one of them emits. The following evolutions of the two electrons are similar to that occurred in the tunneling regime as demonstrated in Figs. 3.10 and 3.11.

3.5.3 Effect of Molecular Alignment

When the light intensity is high enough, it has been the consensus that the DI behavior of atoms is determined by essentially electron physics in the presence of laser field [11, 22]. Good correspondence between our theoretical calculations and experimental data confirms the validity of the above picture in the molecular DI case. In our model, after tunneling, electrons travel much of the time in the intense laser field like a classical object and solely electron collision physics determines the fate of

Fig. 3.13 The molecular alignment dependence of DI ratios for a laser intensity of $0.15P$ W/cm^2. Reprinted with the permission from Ref [29]. Copyright 2007, American Physical Society

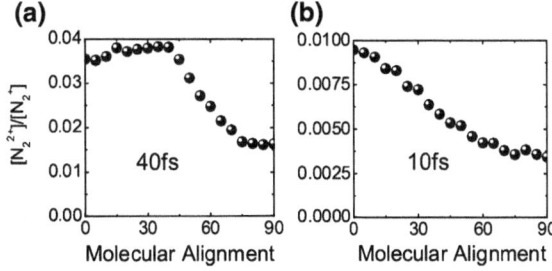

DI of molecules. However, the inherent nuclear degrees of freedom of molecules do manifest themselves as the significant alignment effect in our model. To clearly demonstrate this, we calculate the ratios between double and single ionization at different molecular alignment angles. The main results are presented in Fig. 3.13. It shows that, (i) the ratio between DI and the single-ionization yield is less for perpendicular molecules than that of parallel molecules, and (ii) this anisotropy becomes more dramatic for a shorter laser pulse. Further explorations show that molecular alignment also significantly affects the correlated momentum distribution of emitted electrons [56].

3.6 Double Ionization in Circularly Polarized Laser Fields

It is commonly believed that NSDI will be significantly suppressed in CP fields, because a transverse drift velocity due to the rotating electric field causes the electron to spiral away from the core, prohibiting the recollision and NSDI. This picture has been verified by experiments for helium and xenon [58], not for others such as magnesium [59], NO, and O_2 [60], with the latter showing evidence of NSDI in CP fields. The conflicting evidence between the experiments has motivated some classical simulations [61, 62], which indicate that $\sim 6\%$ of the trajectories are subject to recollision in pure CP fields. The DI yields in the knee regime as calculated from the classical model, however, are one order of magnitude larger than that in experiment [59, 62]. Considering that the NSDI is in the quantum tunneling regime [63], investigation of the NSDI mechanism requires a model beyond the pure classical description. In this section, we investigate double ionization in CP laser fields using our semiclassical quasistatic model.

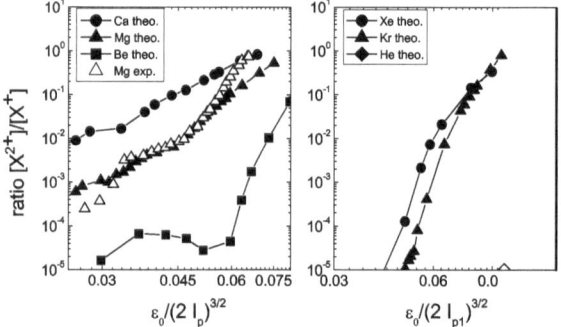

Fig. 3.14 Our model calculation on the ratios of double- over single-ion yield with respect to the scaled field strength. For the alkaline earth metal atoms, apparent knee structures emerge as the signature of NSDI. While for the rare gas atoms, knee structures are absent. Experimental data are from Ref [6]. Reprinted with the permission from Ref [57]. Copyright 2012, American Physical Society

3.6.1 Model Calculation

The ratios of double- over single-ion yields at different field strengths are calculated and shown in Fig. 3.14. The laser wavelength is chosen as 800 nm to match the experiment [64]. As shown in Fig. 3.14, our model calculation agrees with the existing experiments and even reproduces the experimental results for Mg quantitatively. We find that the valence-core interaction described by the screening potentials is important in quantitatively reproducing the experimental data. We compare the result from the screening potential with that from the Coulomb potential, and find that the double ionization yields calculated from the screening potential are larger by one order of magnitude. This is because the screening potential is wider than the Coulomb potential, and the struck electron is readily excited by the recollision.

3.6.2 Correlated Dynamics

To investigate the recollision dynamics, we use a rotating frame (u, v, w) in which the laser field becomes a static electric field in the u direction with constant strength of ε_0. The equations for the outer electron in the rotating frame are:

$$\ddot{u}(t) + 2\omega\dot{v}(t) = -\frac{\partial\Omega}{\partial u}, \tag{3.14}$$

$$\ddot{v}(t) - 2\omega\dot{u}(t) = -\frac{\partial\Omega}{\partial v}, \tag{3.15}$$

$$\ddot{w}(t) = -\frac{\partial\Omega}{\partial w}, \tag{3.16}$$

where, $\Omega = -\frac{1}{r} - \frac{1}{2}\omega^2 r^2 + \varepsilon_0 u$, is the effective potential energy, and $r(t) = \sqrt{u(t)^2 + v(t)^2 + w(t)^2}$.

The electron feels two more forces in addition to Coulomb potential, namely, the centrifugal force $\omega^2 \mathbf{r} = \nabla\frac{1}{2}\omega^2 r^2$, and the Coriolis force $2\omega \times \dot{\sigma}$ ($\omega = \omega\widehat{e}_z$), $\dot{\sigma} = (\dot{u}, \dot{v}, \dot{w})$ is the electron velocity in the rotating frame. The system has a conserved quantity, the so-called Jaccobi integral [65],

$$J = \frac{\dot{\sigma}^2}{2} + \Omega. \tag{3.17}$$

We project potential function Ω onto the plane of (u, v) and plot it in Fig. 3.15a. The parent ion is located at the origin, forming a deep well. The maximum of the potential is located on the positive u axis. On the negative u axis, there is a saddle point, through which the electron can be released via quantum tunneling.

Tracing back the DI trajectories in the knee regime, we find that some tunneled electrons climb up the barrier of effective potential and enter the core regime, interact with the bound electron and successively trigger NSDI, as shown in Fig. 3.15a. In the

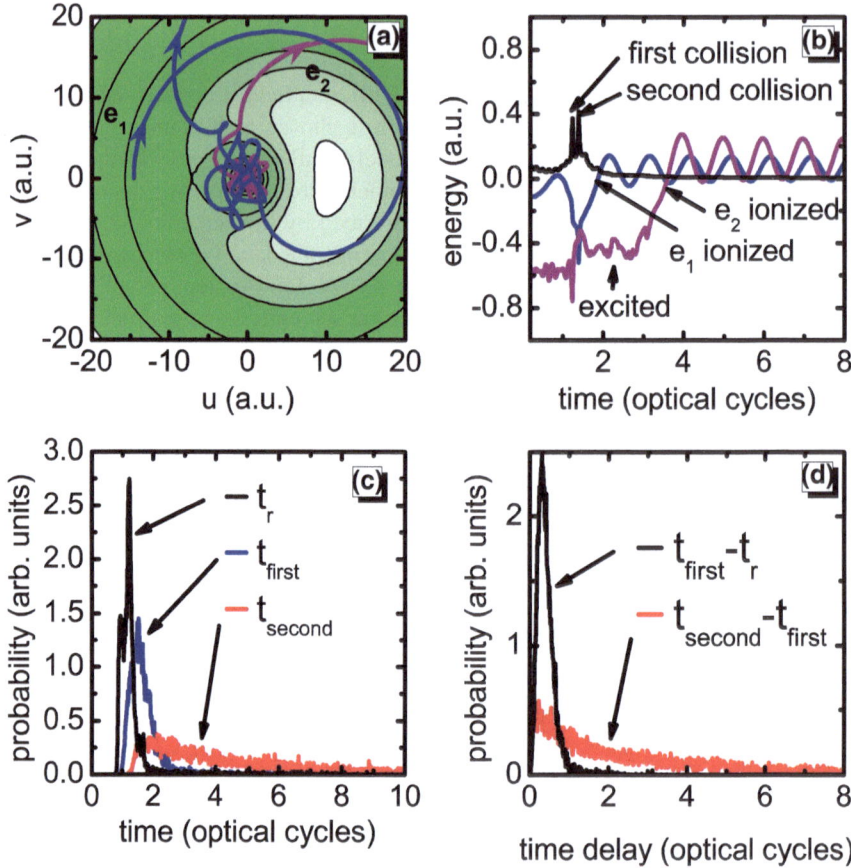

Fig. 3.15 (*Color online*) **a** Contour plot of the potential Ω and typical NSDI trajectories. The tunneled and bound electrons are colored in blue and orange, respectively. **b** Temporal evolution of the electron energies and Coulomb repulsive energy $1/r_{12}$, the latter is colored in black. **c** Statistical analysis on recollision and ionization times. The time zero point in **b** and **c** is moment when the tunneled electron releases. **d** Statistical analysis on time delays. The calculation is made for Mg with the laser parameters of $0.02\,\mathrm{PW/cm^2}$ and $800\,nm$.Reprinted with the permission from Ref [57]. Copyright 2012, American Physical Society

CP case, the recollision is weak because the return energy is usually small and not enough to knock out the inner electron whose bound energy is about 0.55 a.u.. The valence electrons interact with each other through multiple collisions, one electron then emits first, and the other becomes excited and later ionized by the laser field with a long time delay after the recollision (see Fig. 3.15b).

The above picture is more clearly revealed by the calculations on recollision times (t_r), ionization times (Fig. 3.15c) and time delays (Fig. 3.15d). It is shown that the released electrons return to core in about 1.3 optical cycles after tunneling, implying that the electrons will spiral back the third quadrant as shown in Fig. 2.2a. The

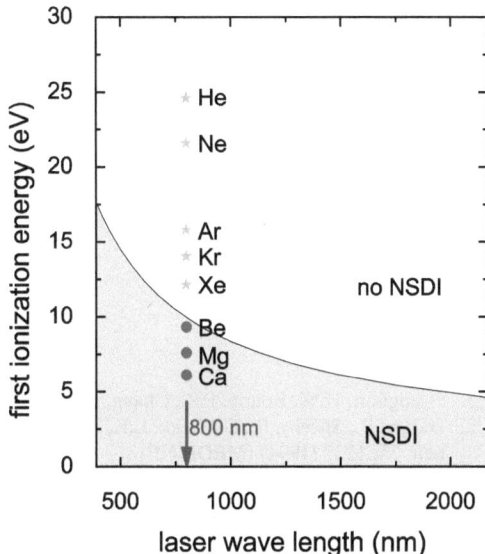

Fig. 3.16 Phase diagram for NSDI in CP fields. The *solid line* is the demarcation line from semiclassical model. The *red color points* indicate that the NSDI signal presents in our semiclassical ensemble simulation while the *green color points* indicate no NSDI sinal presents.Reprinted with the permission from Ref [?]. Copyright 2012, American Physical Society

distribution of the time delays between the ionization of two electrons is much wider, indicating that collision excitation becomes an important channel in the ionization of the second electron. In contrast, the time delay is usually small in LP fields where the collision is strong and DI occurs soon after the recollision [52, 66]. The large time delay trajectories observed here are somewhat similar to the long quantum orbits responsible for double ionization in elliptic polarized fields [63].

3.6.3 Phase Diagramm

According to the semiclassical model, the ratios of double- over single-ion yield can be approximately expressed as, $[X^{2+}]/[X^+] \simeq \frac{1}{10} \frac{\int^{\Delta_{win}} F(v_t)dv_t}{\int_0^\infty F(v_t)dv_t}$, where $F(v_t)$ is the Gaussian distribution on traversal velocity, Δ_{win} is the recollision window, and the prefactor $1/10$ accounts for the fact that only one tenth of the return trajectories can successively trigger NSDI. Consider that the NSDI emerge mainly around critical velocity v_c and the width of the recollision window approximates to ε_0/ω, the above expression reduces to, $[X^{2+}]/[X^+] \simeq \frac{(2I_p^5)^{1/4}}{10\sqrt{\pi}\omega} \exp[-I_p^{5/2}/2\sqrt{2}\omega^2]$. In the above deduction, the typical field strength, i.e., $\varepsilon_0 = \varepsilon_{th}/2 = I_p^2/4$ and long wavelength limit of $\omega \to 0$ are exploited. We then can obtain an explicit expression for the NSDI criterion as $\omega \approx 0.18(I_p)^{5/4}$, that corresponds to a ratio of 10^{-5} for the double to single ionization yield. Below it, the double ionization events are too few to be detectable in the experiments. The analytic result is plotted in Fig. 3.16.

We also put on the rare gas atoms and alkali metal atoms in Fig. 3.16 at an experimental laser wavelength of 800 nm. It is clearly seen that, all the rare gas atoms are above the semiclassical criterion and the alkaline metal atoms are below it, consistent with the existing experiments and our model calculation. The phase diagram exhibits the laser wavelength dependence of NSDI, as confirmed by our numerical simulation. For example, if we shift the laser wavelength to 1, 500 nm our model calculation exhibits that for Mg atom the knee structure disappears.

References

1. Fittinghoff, D.N., Bolton, P.R., Chang, B., Kulander, K.C.: Phys. Rev. Lett. **69**, 2642 (1992)
2. Walker, B., Sheehy, B., DiMauro, L.F., Agostini, P., Schafer, K.J., Kulander, K.C.: Phys. Rev. Lett. **73**, 1227 (1994) [MEDLINE]
3. Sheehy, B., Lafon, R., Widmer, M., Walker, B., DiMauro, L.F., Agostini, P.A., Kulander, K.C.: Phys. Rev. A **58**, 3942 (1998)
4. Corkum, P.B.: Phys. Rev. Lett. **71**, 1994 (1993)
5. Liu, Y.Q., et al.: Phys. Rev. Lett. **101**, 053001 (2008)
6. Weber, Th, Giessen, H., Weckenbrock, M., Urbasch, G., Staudte, A., Spielberger, L., Jagutzki, O., Mergel, V., Vollmer, M., Dö rner, R.: Nature **405**, 658 (2000)
7. Staudte, A., Ruiz, C., Schöffler, M., Schössler, S., Zeidler, D., Weber, Th, Meckel, M., Villeneuve, D.M., Corkum, P.B., Becker, A., Doörner, R.: Phys. Rev. Lett. **99**, 263002 (2007)
8. Fu, L.B., Liu, J., Chen, J., Chen, S.G.: Phys. Rev. A **63**, 043416 (2001)
9. Landau, L.D., Lifishitz, E.M. (ed.): Quantum mechanics. Pergamon, Oxford (1977)
10. Abrines, R., Percival, I.C.: Proc. Phys. Soc. London **88**, 861 (1966)
11. Leopold, J.G., Percival, I.C.: J. Phys. B **12**, 709 (1979)
12. Cohen, J.S.: Phys. Rev. A **26**, 3008 (1982)
13. Staudte, A., et al.: Phys. Rev. Lett. **99**, 263002 (2007)
14. Rudenko, A., et al.: Phys. Rev. Lett. **99**, 263003 (2007)
15. de Morisson Faria, C.F., et al.: Phys. Rev. A **69**, 043405 (2004); ibid **69**
16. de Morisson Faria C.F., et al.: J. Phys. B **38**, 3251 (2005)
17. Parker, J.S., et al.: Phys. Rev. Lett. **96**, 133001 (2006)
18. Prauzner-Bechcicki, Jakub S., et al.: Phys. Rev. Lett. **98**, 203002 (2007)
19. Milosevic, D., Becker, W.: Phys. Rev. A **68**, 065401 (2003)
20. Ye, D.F., Liu, X., Liu, J.: Phys. Rev. Lett. **101**, 233003 (2008)
21. Berakdar, J., et al.: J. Phys. B **29**, 6203 (1996)
22. Ho, Phay J., et al.: Phys. Rev. Lett. **94**, 093002 (2005)
23. The recollision time is defined as the instant when the two electrons get closest, while the DI time as that when both electron energy become greater than zero.
24. Haan, S.L., et al.: Phys. Rev. Lett. **97**, 103008 (2006)
25. Paulus, G.G., et al.: J. Phys. B **27**, L703 (1994)
26. Weckenbrock, M., et al.: Phys. Rev. Lett. **92**, 213002 (2004)
27. Feuerstein, B., et al.: Phys. Rev. Lett. **87**, 043003 (2001)
28. Liu, X., et al.: J. Phys. B **39**, L305 (2006)
29. Liu, J., et al.: Phys. Rev. Lett. **99**, 013003 (2007)
30. Weber, Th, Giessen, H., Weckenbrock, M., et al.: Nature (Lond.) **405**, 658 (2000)
31. Staudte, A., Ruiz, C., Schöffler, M., et al.: Phys. Rev. Lett. **99**, 263002 (2007)
32. Rudenko, A., de Jesus, V.L.B., Ergler, T., et al.: Phys. Rev. Lett. **99**, 263003 (2007)
33. Haan, S.L., Smith, Z.S., Shomsky, K.N., Plantinga, P.W.: J. Phys. B **41**, 211002 (2008)
34. Bondar, D.I., Liu, W.-I.: Ivanov. Phys. Rev. A **79**, 023417 (2009)
35. Fu, L.-B., Liu, J., Chen, J., Chen, S.-G.: Phys. Rev. A **63**, 043416 (2001)

36. Fu, L.-B., Liu, J., Chen, S.-G.: Phys. Rev. A **65**, 021406(R) (2002)
37. Ye, D.-F., Liu, J.: Phys. Rev. A **81**, 043402 (2010)
38. Cohen, J.S.: Phys. Rev. A **64**, 043412 (2001)
39. Dimitriou, K.I., Arbó, D.G., Yoshida, S., Persson, E., Burgdörfer, J.: Phys. Rev. A **70**, 061401 (2004)
40. Guo, C., Li, M., Nibarger, J.P., Gibson Phys, G.N.: Rev. A **58**, R4271 (1998)
41. Cornaggia, C., Hering, Ph: Phys. Rev. A **62**, 023403 (2000)
42. Muth-Böhm, J., Becker, A., Faisal, F.H.M.: Phys. Rev. Lett. **85**, 2280 (2000)
43. Kjeldsen, T.K., Madsen, L.B.: J. Phys. B: At. Mol. Opt. Phys. **37**, 2033 (2004)
44. Alnaser, A.S., Voss, S., Tong, X.M., et al.: Phys. Rev. Lett. **93**, 113003 (2004)
45. Eremina, E., Liu, X., Rottke, H., et al.: Phys. Rev. Lett. **92**, 173001 (2004)
46. Zeidler, D., Staudte, A., Bardon, A.B., Villeneuve, D.M., Dörner, R., Corkum, P.B.: Phys. Rev. Lett. **95**, 203003 (2005)
47. Parker. J.S., et al.: J. Phys. B **34**(3), L69–L78 (2001)
48. Pegarkov, A.I., Charron, E., Suzor-Weiner, A.: J. Phys. B **32**, L691 (2000)
49. Becker, A., Faisal, F.H.M.: J. Phys. B **38**, R1 (2005)
50. Prauzner-Bechcicki, Jakub S., Sacha, Krzysztof, Eckhardt, Bruno, Zakrzewski, Jakub: Phys. Rev. A **71**, 033407 (2005)
51. The atomic ADK theory has been extended to diatomic molecules; see, for example, X. M. Tong et al., Phys. Rev. A **66** 033402 (2002), and I. V. Litvinyuk et al., Phys. Rev. Lett. **90**, 233003 (2003). However, we found that the employment of atomic ADK formula instead of the complicated molecular ADK formula does not lead to significant discrepancy in calculating the ratios between double and single ionization. So, for simplicity, we adopt $\varpi(t_0) = \frac{4(2I_{p1})^2}{\varepsilon(t_0)} \exp(-\frac{2(2|I_{p1}|)^{3/2}}{3\varepsilon(t_0)})$ in our calculations
52. Ye, D.F.., et al.: Phys. Rev. A **77**, 013403 (2008)
53. Prauzner-Bechcicki, Jakub S., Sacha, Krzysztof, Eckhardt, Bruno, Zakrzewski, Jakub: Phys. Rev. A **71**, 033407 (2005)
54. Meng, L., Reinhold, C.O., Olson, R.E.: Phys. Rev. A **40**, 3637 (1989)
55. Eichenauer, D., Grün, N., Scheid, W.: J. Phys. B **14**, 3929 (1981)
56. Li, Y., Chen, J., Yang, S.P., Liu, J.: Phys. Rev. A **76**, 023401 (2007)
57. Fu, L.B., Xin, G.G., Ye, D.F., Liu, J.: Phys. Rev. Lett **108**, 103601 (2012)
58. Walker, B., et al.: Phys. Rev. A **48**, R894 (1993)
59. Gillen, G.D., Walker, M.A., VanWoerkom, L.D.: Phys. Rev. A **64**, 043413 (2001)
60. Guo, C., Gibson, G.N.: Phys. Rev. A **63**, 040701(R) (2001)
61. Wang, X., Eberly, J.H.: Phys. Rev. Lett. **105**, 083001 (2010)
62. Mauger, F., Chandre, C., Uzer, T.: Phys. Rev. Lett. **105**, 083002 (2010)
63. Shvetsov-Shilovski, N.I., Goreslavski, S.P., Popruzhenko, S.V., Becker, W.: Phys. Rev. A **77**, 063405 (2008)
64. Gillen, G.D., Walker, M.A., Vanwoerkom, L.D.: Phys. Rev. A. **64**, 043413 (2001)
65. Szebehely, V.G.: In the retsricted three-body problem: the field term is replaced by the gravity potential of the second primary body. Theory of Orbits. Academic, New York (1967)
66. Haan, S.L., et al.: Phys. Rev. Lett. **97**, 103008 (2006)

Chapter 4
Partition of the Linear Photon Momentum in Atomic Tunneling Ionization

Abstract In this chapter, partitioning of photon momentum between the ion and electron in photoionization and the involved subcycle dynamics are investigated using an extended semiclassical model, where momentum transfer from the photon to the electron ion system via the (magnetic) Lorentz force is taken into account.

4.1 Brief Introduction

In light-atom interaction, besides the often invoked photon energy and angular momentum [1, 2], another property, namely the linear photon momentum, is seldom considered. This is partly because the linear (hereafter omitted for simplicity) momentum of a visible photon is extremely small. On the other hand, an intense laser pulse contains many photons and their combined momentum can give rise to macroscopic effects [3]. The latter include radiation pressure as proposed by Gigan et al. [4], which has important applications such as manipulation of cold atoms [5], acceleration of particles [6, 7], generation of terahertz radiation [8–11], etc.

An intriguing question associated with photon momentum transfer is: in strong-field atomic ionization, how is the photon momentum partitioned between the electron and ion? In this chapter, we propose an extended semiclassical model for photon momentum partitioning between the ion and electron in atomic ionization by intense laser light [12]. The model includes tunneling ionization and the full dynamics of the valence electron and ion in the combined laser and Coulomb fields (e.g. electron-core system). It can thus also be used to investigate the subcycle dynamics of the photon momentum partitioning process.

J. Liu, *Classical Trajectory Perspective of Atomic Ionization in Strong Laser Fields*,   59
SpringerBriefs in Physics, DOI: 10.1007/978-3-642-40549-5_4, © The Author(s) 2014

4.2 Semiclassical Model for Partition of Photon Momentum

For the clarity, in this step we assume the electric field to be quasistatic, and place it in the z direction. The tunneling rate associated with electron tunneling at time t_0 is $w(t_0) = \left(\frac{\varepsilon(t_0)}{4}\right)\left(\frac{4\kappa^4}{\varepsilon(t_0)}\right)^{2/\kappa} e^{-\frac{2\kappa^3}{3\varepsilon(t_0)}}$, in which $\kappa = \sqrt{2I_p}$ and $\varepsilon = \varepsilon(t_0)$ is the instantaneous electric field.

The electron after tunneling has a velocity distribution in the plane perpendicular to the instantaneous laser electric field as $f(v_x, v_y) = \frac{\kappa}{\pi\varepsilon(t_0)} e^{-\kappa\left((v_{x0}^e)^2 + (v_{y0}^e)^2\right)/\varepsilon(t_0)}$ [13], so the distribution of the magnitude of the velocity $v_{0\perp}^e$ is $f(v_{0\perp}^e) = \frac{2\kappa v_{0\perp}^e}{\varepsilon} e^{\frac{-\kappa(v_{0\perp}^e)^2}{\varepsilon}}$. In practical calculation, we choose random $v_{0\perp}^e$ in a proper range and random azimuthal angle in the interval $[0, 2\pi]$, associated with a rate $f(v_\perp)$ [14].

As for the initial position and velocity of ion core, it is reasonable to place it at the origin at first. The energy increment of the electron-core system in the tunneling process equals $E_{k0}^e - 1/r_0 + I_p$. In the above expression $E_{k0}^e = \frac{1}{2}v_\perp^2$ is the energy of the electron after tunneling, $r_0 = \sqrt{x_0^2 + y_0^2 + z_0^2}$, and we ignore the ion kinetic energy because it is negligibly small compared to both the electric energy and ionization potential. Since the momentum of the photons absorbed by the atom must be transferred to the system at the same time, the net momentum gain of electron ion system ensures the relation $\frac{E_{k0}^e - 1/r_0 + I_p}{c} = p_{z0}^e + p_{z0}^i$ in the tunneling process. We then properly assign initial velocities to ion according to

$$Mv_{x0}^i = -mv_{x0}^e, \quad Mv_{y0}^i = -mv_{y0}^e, \tag{4.1}$$

$$Mv_{z0}^i = \frac{E_{k0}^e - 1/r_0 + I_p}{c} - mv_{z0}^e, \tag{4.2}$$

in which M and m are the mass of ion and electron, respectively.

In the above, we have obtained the initial condition of the electron and the ion core after tunneling. Next the motion of the whole is governed by the classical Newton's law.

We denote by $\mathbf{r}^i$, $\mathbf{r}^e$ the coordinates and $\mathbf{v}^i$, $\mathbf{v}^e$ the velocities of the ion core and the electron, respectively, and the Newton's equation of motion for the ion and the electron can be written as

$$M\frac{d^2\mathbf{r}^i}{dt^2} = \frac{\mathbf{r}^e - \mathbf{r}^i}{\left(|\mathbf{r}^e - \mathbf{r}^i|^2\right)^{3/2}} + e\mathbf{E}\left(\mathbf{r}^i\right) + e\mathbf{v}^i \times \mathbf{B}\left(\mathbf{r}^i\right)/c, \tag{4.3}$$

$$m\frac{d^2\mathbf{r}^e}{dt^2} = -\frac{\mathbf{r}^e - \mathbf{r}^i}{\left(|\mathbf{r}^e - \mathbf{r}^i|^2\right)^{3/2}} - e\mathbf{E}\left(\mathbf{r}^e\right) - e\mathbf{v}^e \times \mathbf{B}\left(\mathbf{r}^e\right)/c, \tag{4.4}$$

in which $\mathbf{E}$, $\mathbf{B}$ are the electric and magnetic components of the laser field.

The electric field is given by

$$\mathbf{E} = \frac{\varepsilon_0}{\sqrt{\chi^2 + 1}} \exp\left[-\frac{2\ln 2}{c^2\tau^2}(z - ct)^2\right]$$
$$\times \left(\cos(\omega t - kz)\,\mathbf{e}_x + \chi \sin[\omega t - kz]\,\mathbf{e}_y\right) \tag{4.5}$$

where χ is the ellipticity. The magnetic field is given by $\mathbf{e}_z \times \mathbf{E}/c$. For each t_0 we can calculate the electron tunneling point and the electron velocity $v_\perp$ perpendicular to instantaneous electric field, and attribute a rate $w(t_0)\, f(v_\perp)$ to the event. We then take average overall velocities and tunneling phases to obtain the physical quantities of interest, such as momentum gain and loss of the electron and ion (see Chap. 5).

4.3 Coulomb Attraction Effects

We apply the model to Ne atom with the ionization potential parameters $I_p = 0.79$ a.u. The screening potential is introduced to depict the interaction between valence electron and ion, ie,

$$V_s(\mathbf{r}_i - \mathbf{r}_e) = -[(Z - 1)s(|\mathbf{r}_i - \mathbf{r}_e|) + 1]/|\mathbf{r}_i - \mathbf{r}_e|, \tag{4.6}$$

where $s(r) = [H(e^{r/d} - 1) + 1]^{-1}$, Z is the nuclear charge, and H and d are two atomic parameters [15]. The wavelength of laser light is 800 nm and the pulse duration is $\tau = 15$ ms [16]. In our simulations, the Newton's equations are solved using a 4–5th Runge Kutta numerical algorithm. More than 5×10^6 classical trajectories are used and numerical convergence has been tested by doubling the number of trajectories.

We present in Fig. 4.1a (and black dotted line in c) the results for CP light. In Fig. 4.1b, one can see two bright spots in the spectrum, which we attribute to rescattering in LP light. The blue dotted line in Fig. 4.1c clearly reveals Coulomb focusing [17]. Both distributions in Fig. 4.1c show net momentum shifts $p_z > 0$ along the propagation direction.

Figure 4.2 shows the electron and ion averaged momentum (i.e., net momentum gain) as a function of the light intensity. From Fig. 4.2a, we see that our model calculations on the electron momentum gains in CP fields are in good agreement with experimental observations [16]. The net electron momentum was also be obtained by the quantum scattering matrix without considering the ion's motion [18]. In contrast to the quantum treatment, in our model, the electron gains or loses longitudinal momentum on its way out of the laser beam is automatically and fully included. Our simulations at the same time are able to trace the motions of ions and generate the ion momentum gains as shown in Fig. 4.2b. For CP fields, the averaged ion momentum p_z^i fairly follows the I_p/c dependence with small deviations that depend on the laser intensity. While, for the LP case, we find that the net ion momentum exhibits a sudden increment around 2×10^{14} W cm^{-2}, above which the simulation

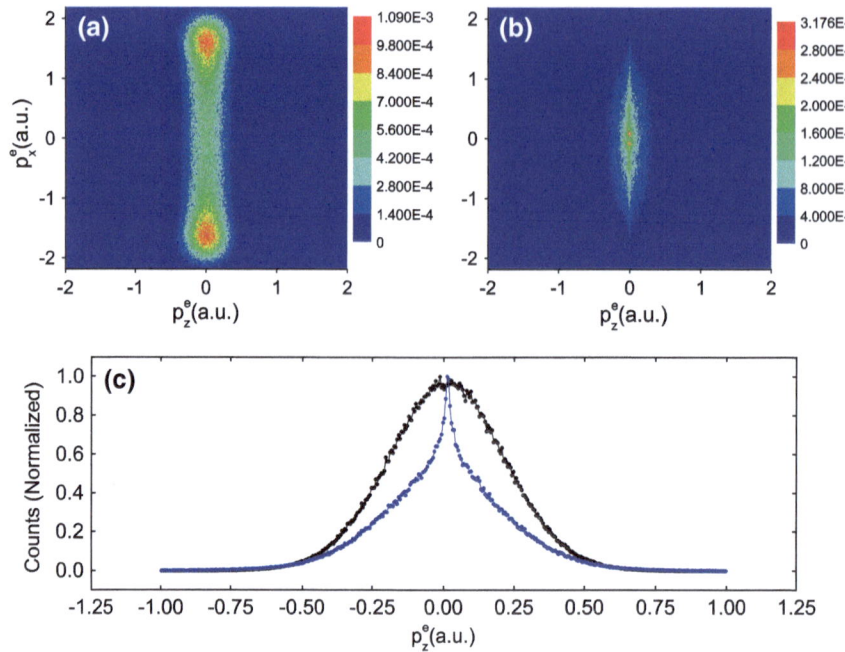

Fig. 4.1 Simulated electron-momentum spectrum for Ne atom ionization by 800 nm, 8×10^{14} W/cm^2 lasers of CP (**a**) and LP (**b**) polarization. **c** Simulated Ne photoelectron momentum distribution along the propagation direction. *Black/blue*: results for the CP/LP light. Reprinted with the permission from Ref. [30]. Copyright 2013, American Physical Society

results deviate from the assertion [16, 18] $p_z^i = I_p/c$ by up to 30 %. We have also checked the momentum-energy relation $p_z^e + p_z^i = (E_k^e + I_p)/c$ from our simulation. Note that, even with inclusion of Coulomb interaction, our model does not produce negative [16] averaged electron momentum at low intensities.

The "simple-man model" [19–21] of atomic ionization invokes the relation $p_z^i = I_p/c$ or $p_z^e = E_k^e/c$, which assumes that the electron is tunneled out with zero initial velocity at the origin and then executes quiver motion in the light wave. In the tunneling process, the electron and ion are tightly bounded, and the part of the longitudinal momentum I_p/c, corresponding to the photon momentum necessary to overcome the ionization energy, must be transferred to the center of the mass of the electron ion system, i.e., $p_z^i = I_p/c$. In the post-tunneling process, the electron absorbs excessive photon energy and momentum, so that one obtains $p_z^e = E_k^e/c$. This simple picture ignores the Coulomb attraction between the ion and electron in both the tunneling and post-tunneling scattering processes, therefore can not account for the above simulations.

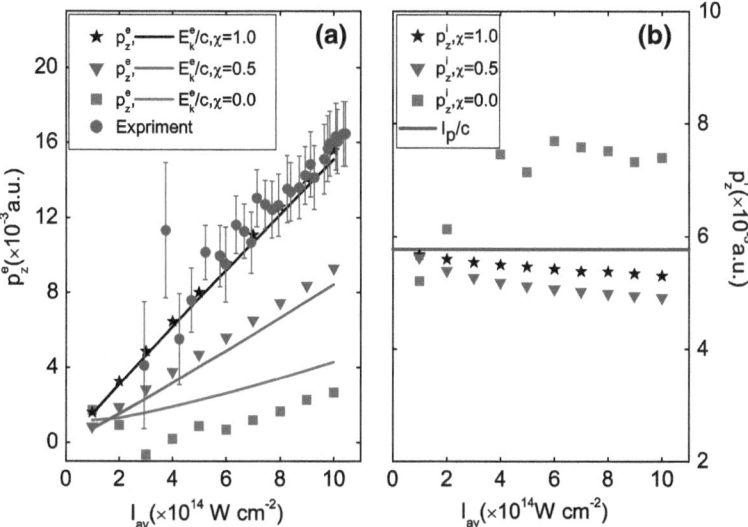

Fig. 4.2 **a** Averaged electron momentum along the propagation (z) direction as a function of the laser intensity. The *dots* denote the simulated averaged electron momentum. The *lines* denote the simulated averaged electron kinetic energy E_k^e divided by the vacuum light speed c. *Blue*: experimental data from Ref. [16]. *Black*: results for CP light with $\chi = 1$. *Green*: results of elliptically polarized light with $\chi = 0.5$. *Red*: results for LP light with $\chi = 0$. **b** Averaged ion momentum along the laser propagation direction. The (*violet*) horizontal line is for $p_z^i = I_p/c$. Reprinted with the permission from Ref. [30]. Copyright 2013, American Physical Society

4.4 Classical Trajectory Analysis

Atomic ionization including Coulomb interaction can be described by an effective potential in parabolic coordinates [22]. The tunneled electron is emitted along the direction of instantaneous electric field at a nonzero distance of r_0 from the ion [23]. The distance depends on the field strength and the ionization potential and can be approximated by $r_0 \simeq I_p/\varepsilon$. In addition, the tunneled electron has a Gaussian transverse velocity distribution of $v_\perp$. That is, before tunneling, the electron ion system has a bound energy of $-I_p$, and after tunneling, the system energy is $v_\perp^2/2 - 1/r_0$, so that the system energy increase in the tunneling process is $v_\perp^2/2 - 1/r_0 + I_p$. Here we have ignored the kinetic energy of the ion. The photon momentum should be shared between the ion and electron, that is,

$$p_{z0}^e + p_{z0}^i = (v_\perp^2/2 - 1/r_0 + I_p)/c. \tag{4.7}$$

Taking average over the ensemble of all transverse velocities and noting that $\langle p_{z0}^e \rangle = 0$, $\langle v_\perp^2 \rangle = \varepsilon/\sqrt{2I_p}$, we get

$$\langle p_{z0}^i \rangle = \left(\varepsilon/2\sqrt{2I_p} - \varepsilon/I_p + I_p \right)/c, \tag{4.8}$$

where $\varepsilon = \varepsilon_0 \exp[-2\ln 2(t_0^2/\tau^2)]$ is the electric field at t_0 in the plane $z = 0$. In contrast to the "simple-man model", the net longitudinal momentum acquired by the ion in the tunneling process deviates from I_p/c by a small quantity of order ε_0, for both LP and CP laser fields (see Sect. 4.2).

The Coulomb attraction between the ion and electron is also involved in the post-tunneling scattering process. In CP laser fields, the tunneled electron has a velocity proportional to the instantaneous electric field strength and is directed perpendicular to the instantaneous polarized field direction. The electron spirals away from its parent ion without returning [24]. Thus, the ion can only acquire longitudinal momentum from a limited number of photons. In the LP fields, on the other hand, the electron in a certain initial phase window can revisit its parent ion [14, 25, 26], leading to the re-partitioning of the photon momentum between the ion and electron. The momentum transfer to the ion is found to be significantly reduced.

The scenario given above is helpful for understanding the observed sudden change of the net ion momentum at the laser intensity 2×10^{14} W cm^{-2} in Fig. 4.2b. The tunneled electron is released at a distance I_p/ε_0 from the core ion. The characteristic displacement of the quiver motion of the tunneled electron in the electric field is ε_0/ω^2. With the decrease of the laser intensity, the quiver length decreases while the exit distance increases. When the quiver length is smaller than the exit distance, the electron will have little chance to recollide with the ion. The corresponding threshold intensity is $\varepsilon_0^{th} \simeq \omega\sqrt{I_p}$, which approximates to be 1×10^{14} W cm^{-2}, comparable with the numerical results. The deviation is due to the electron traverse velocity that might decrease the collision possibility and increase the threshold field.

Figure 4.3a–c shows three pairs of typical electron trajectories corresponding to three different initial phase regimes, i.e., $\omega t_0 < -0.2$, $-0.2 < \omega t_0 < 0.1$ and $\omega t_0 > 0.1$, respectively. The trajectories are projected on the plane of polarization direction (x) and light propagation direction (z). For each pair of electron orbits, the initial velocities are opposite but have the same amplitude. Note that the two orbits are asymmetric with respect to the x axis because the photons can push the electrons forward along the propagation direction. These electrons can in turn pull the ion core forward along the positive z direction through the Coulomb attraction force. Figure 4.3a also shows that the electrons move out directly, with some quiver oscillations. Figure 4.3c shows the electron trajectories for $\omega t_0 > 0.1$, where the electron can return to the neighborhood of the ion core once. We present in Fig. 4.3 d–f the time resolved net ion momentum. The behavior for the initial phases $\omega t_0 = -0.5$ and 0.5 are relatively simple: both show almost monotonous increase initially and then saturation when the electrons are far away from the ion. The increase for $\omega t_0 = 0.5$ is larger because, in this case, the net ion momentum gain mainly emerges when the electrons return to ion and at this moment, the electrons are closer to ion and can pull the ion forward more strongly.

When the initial phase is in the regime $-0.2 < \omega t_0 < 0.1$, the post-tunneling electron trajectories are complex. Figure 4.3b shows that multiple returns and collisions

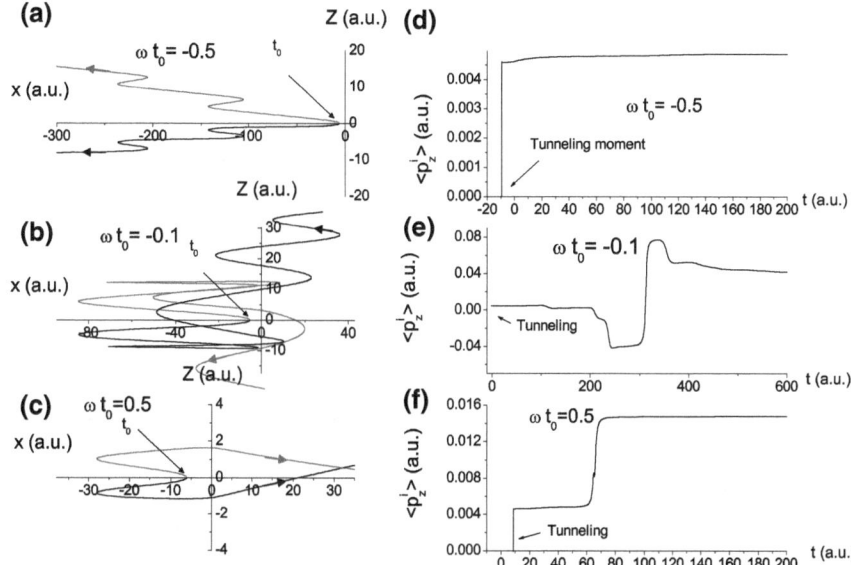

Fig. 4.3 Typical electron trajectories with different tunneling phases of **a** −0.5, **b** −0.1 and **c** 0.5. For each phase, a pair of trajectories are plotted with initial opposite velocities. **d–f** The time resolved net ion momentum averaged over the trajectory pairs corresponding to **a–c**, respectively. Reprinted with the permission from Ref. [30]. Copyright 2013, American Physical Society

with the ion core can occur. We see that Coulomb attraction between the returning electrons and the ion core at collisions leads to a dramatic change in the orbits, which in turn modifies the momentum transfer. For $\omega t_0 = -0.1$, the net ion momentum first decreases, then increase and finally decrease with some step structures in between. The electrons can long-term entangle with ion, weakly bounded by Coulomb attraction. At this stage, the photon prefer to transfer its momentum to ion rather than electron and the ion can obtains a large amount of longitudinal momenta on average. It is of interest to point out that if the electron tunnels at $\omega t_0 \sim -0.1$, the final ion momentum is very sensitive to the electron's initial velocity, indicating onset of chaos [14, 27]. Chaotic orbits have been shown to account for the higher above-threshold energy spectra [14], high-order harmonic generation [28] and double ionization [24, 29]. It is evident that chaos is involved in the photon momentum partitioning.

References

1. Gavrila, M.: Atoms in Intense Laser Fields. Academic Press, San Diego (1992)
2. Brabec, T.: Strong Field Laser Physics. Springer, New York (2009)
3. Marx, G.: Nature (London) **211**, 22 (1966)
4. Gigan, S., et al.: Nature (London) **444**, 67 (2006)
5. Chin, C., Grimm, R., Julienne, P., Tiesinga, E.: Rev. Mod. Phys. **82**, 1225 (2010)

6. Eichmann, U., et al.: Nature **461**, 1261 (2009)
7. Hegelich, M., et al.: Physical Review Letters **89**, 085002 (2002)
8. Amico, C.D., et al.: New J. Phys. **10**, 013015 (2008)
9. Sprangle, P., Penano, J.R., Hafizi, B., Kapetanakos, C.A.: Phys. Rev. E **69**, 066415 (2004)
10. Cheng, C.C., Wright, E.M., Moloney, J.V.: Physical Review Letters **87**, 213001 (2001)
11. Zhou, B., Houard, A., Liu, Y., Prade, B., Mysyrowicz, A., Couairon, A., Mora, P., Smeenk, C., Arissian, L., Corkum, P.: Physical Review Letters **106**, 255002 (2011)
12. Liu, J., Xia, Q.Z., Tao, J.F., Fu, L.B.: Phys. Rev. A **87**, 041403 (2013)
13. Delone, N.B., Krainov, V.P.: J. Opt. Soc. Am. B **8**, 1207 (1991)
14. Hu, B., Liu, J., Chen, S.G.: Phys. Lett. A **236**, 533 (1997)
15. Szydlik, P.P., Kutcher, G.J., Green, A.E.S.: Physical Review A **10**, 1623 (1974)
16. Smeenk, C.T.L., Arissian, L., Zhou, B., Mysyrowicz, A., Villeneuve, D.M., Staudte, A., Corkum, P.B.: Phys. Rev. Lett. **106**, 193002 (2011)
17. Brabec, T., Ivanov, M.Y., Corkum, P.B.: Phys. Rev. A **54**, 2551 (1996)
18. Titi, A.S., Drake, G.W.F.: Phys. Rev. A. **85**, 041404 (2012)
19. van Linden, H.B., Muller, H.G.: In: Smith, S.J., Knight, P.L. (eds.) Multiphoton Processes. Cambridge University Press, Cambridge (1988)
20. Gallagher, T.F.: Phys. Rev. Lett. **61**, 2304 (1988)
21. Corkum, P.B., Burnett, N.H., Brunel, F.: Physical Review Letters **62**, 1259 (1989)
22. Landau, L.D., Lifshitz, E.M.: Quantum Mechanics. Pergamon, New York (1977)
23. Hickstein, D.D., Ranitovic, P., Witte, S., Tong, X.M., et al.: Phys. Rev. Lett. **109**, 073004 (2012)
24. Fu, L.B., Xin, G.G., Ye, D.F., Liu, J.: Phys. Rev. Lett. **108**, 103601 (2012)
25. Corkum, P.B.: Phys. Today **64**, 36 (2011)
26. Corkum, P.B.: Phys. Rev. Lett. **71**, 1994 (1993)
27. Liu, J., Chen, S.G., Hu, B.: Acta Sin Phys **7**, 89 (1998)
28. van de Sand, G., Rost, J.M.: Phys. Rev. Lett. **83**, 524 (1999)
29. Mauger, F., Chandre, C., Uzer, T.: Phys. Rev. Lett. **102**, 173002 (2009)
30. Liu, J., Xia, Q.Z., Tao, J.F., Fu, L.B.: Physical Review A **87**, 041403 (2013)

Chapter 5
Acceleration of Neutral Atoms with Polarized Intense Laser Fields

Abstract We investigate the dynamics of accelerating neutral atoms with linearly or circularly intense laser pulses. Our classical-trajectory ensemble simulations involve all nuclear and electronic dynamics including tunneling ionization and scattering of electrons in the combined Coulomb and electromagnetic fields for both helium (He) and magnesium (Mg).

5.1 Brief Introduction

Intense laser-based particle acceleration has attracted much attention in recent years due to their important applications in physics as well as in many other fields, such as medical science [1–3]. High-energetic ion (up to a few 100 Mev) [4–6] or electron (up to a few Gev) [7, 8] beams have been achieved and many possible applications have been discussed [9, 10]. In contrast to the acceleration of these charged particles, discussion on acceleration of neutral atoms and molecules is relatively few. Nevertheless, acceleration of neutral atoms or molecules is of great importance in both fundamental [11] and applied physics, such as controlled collision [12–15], atomic nanofabrication [16, 17] and atom optics [18]. The difficulty in neutral atom acceleration is the high ionization probability [19, 20] induced by strong laser fields. Recent experiments, however, find that significant amount of atoms can survive after the laser pulse, showing the phenomenon of tunneling without ionization [21, 22]. Moreover, the concept of partial atomic stabilization in tunneling ionization is proposed and analyzed thoroughly both experimentally and theoretically [23]. The above investigations indicate that neutral atoms or molecules might be effectively accelerated in laser fields. In fact, in the recent experiment of Eichmann et al. [24], up to 10^{14} g (g: the Earth's gravitational acceleration) of acceleration for He atom is obtained using linearly polarized (LP) lasers of peak intensities around 10^{16} w/cm^2.

The dynamics behind neutral atom acceleration is rather complex. It involves the electron tunneling process dominated by the interplay of the Coulomb force and electric field force and the post-tunneling dynamics of the nucleus and electron [25]. In the post-tunneling process the charged particles exposed to an oscillating electromagnetic fields will experience a force proportional to the cycle-averaged intensity gradient, i.e., ponderomotive force [26]. Thus, the complex dynamics of the neutral atom acceleration is far from fully understood while it is of particular interest because of its important implications in both theoretical and applied physics.

In the following, we develop a classical-trajectory Monte Carlo method with allowance of tunneling (CTMC+T) to account for the complex dynamics of neutral atom acceleration [27].

5.2 Extended CTMC+T Model for Atomic Acceleration

We first extend the classical-trajectory Monte Carlo method with allowance of tunneling (CTMC+T) to account for the complex dynamics of the neutral atom acceleration. The original CTMC+T model [28] is applicable in simulating the hydrogen atom ionization process in intense laser fields, in which the nuclear motion is frozen. In order to elucidate the acceleration of atoms, we extend the model to include the motion of nucleus.

The initial coordinate $\mathbf{r}_{e0}$ and velocity $\dot{\mathbf{r}}_{e0}$ of valence electron satisfies the microcanonical distribution $f(\mathbf{r}_{e0}, \dot{\mathbf{r}}_{e0})$ [29],

$$f(\mathbf{r}_{e0}, \dot{\mathbf{r}}_{e0}) = \frac{\delta(I_p - H_0(\mathbf{r}_{e0}, \dot{\mathbf{r}}_{e0}))}{K}, \tag{5.1}$$

where $H_0(\mathbf{r}_{e0}, \dot{\mathbf{r}}_{e0}) = \dot{\mathbf{r}}_{e0}^2/2 - 1/|\mathbf{r}_{N0} - \mathbf{r}_{e0}|$, K is the normalization constant, I_p is the ionization potential, and $\mathbf{r}_{N0}$ denotes the initial position of nucleus. Throughout the paper atomic units are used, i.e., $e = \hbar = m_e = 1$, unless otherwise specified. The initial velocity $\dot{\mathbf{r}}_{N0}$ of nucleus meets the requirement that the total initial momentum of the atom is zero, i.e., $\dot{\mathbf{r}}_{e0} + m_N \dot{\mathbf{r}}_{N0} = 0$, so that the final velocity of atom after pulse is utterly originated from laser field acceleration. Given the initial conditions presented above the subsequent evolutions of the electron and nucleus are governed by the Newton's equations

$$\ddot{\mathbf{r}}_N = \frac{1}{m_N}(\frac{\mathbf{r}_e - \mathbf{r}_N}{|\mathbf{r}_N - \mathbf{r}_e|^3} + \mathbf{E}(\mathbf{r}_N, t) + \dot{\mathbf{r}}_N \times \mathbf{B}(\mathbf{r}_N, t)), \tag{5.2}$$

$$\ddot{\mathbf{r}}_e = \frac{\mathbf{r}_N - \mathbf{r}_e}{|\mathbf{r}_N - \mathbf{r}_e|^3} - \mathbf{E}(\mathbf{r}_e, t) - \dot{\mathbf{r}}_e \times \mathbf{B}(\mathbf{r}_e, t). \tag{5.3}$$

The electrical field is expressed as follows:

$$\mathbf{E}(\mathbf{r}, t) = E_x(\mathbf{r}, t)\hat{e}_x + E_y(\mathbf{r}, t)\hat{e}_y, \tag{5.4}$$

and

$$E_x(\mathbf{r}, t) = \frac{E_0(\mathbf{r})}{\sqrt{1 + \chi^2}} \cos(\omega t - kz), \tag{5.5}$$

$$E_y(\mathbf{r}, t) = \frac{\chi E_0(\mathbf{r})}{\sqrt{1 + \chi^2}} \sin(\omega t - kz), \tag{5.6}$$

where χ is the laser polarization ellipticity, $\chi = 0$ and $\chi = 1$ correspond to LP and Circularly Polarized (CP), respectively. $E_0(\mathbf{r})$ can be written in cylindrical coordinates as

$$E_0(\mathbf{r}) = \epsilon_0 e^{-\frac{r^2}{r_0^2}} (1 + \frac{z^2}{z_0^2})^{-\frac{1}{2}}. \tag{5.7}$$

where

$$z_0 = \pi w_0^2 / \lambda, \tag{5.8}$$

$$r_0 = w_0 \sqrt{1 + (z/z_0)^2}, \tag{5.9}$$

where, w_0 is the laser beam waist, ϵ_0 is the slowly-varied envelop of electric field strength. According to Faraday's law $-\partial \mathbf{B}(\mathbf{r}, t)/\partial t = \nabla \times \mathbf{E}(\mathbf{r}, t)$, the magnetic field takes the form

$$\begin{aligned} B_x(\mathbf{r}, t) = &-\frac{E_0(\mathbf{r})k\chi \sin(\omega t - kz)}{\omega\sqrt{1 + \chi^2}} + \frac{E_0(\mathbf{r})\chi z \cos(\omega t - kz)}{\omega z_0^2(1 + z^2/z_0^2)\sqrt{1 + \chi^2}} \\ &- \frac{2E_0(\mathbf{r})r^2\chi z \cos(\omega t - kz)}{\omega r_0^2 z_0^2(1 + z^2/z_0^2)\sqrt{1 + \chi^2}}, \end{aligned} \tag{5.10}$$

$$\begin{aligned} B_y(\mathbf{r}, t) = &\frac{E_0(\mathbf{r})k \cos(\omega t - kz)}{\omega\sqrt{1 + \chi^2}} + \frac{E_0(\mathbf{r})z \sin(\omega t - kz)}{\omega z_0^2(1 + z^2/z_0^2)\sqrt{1 + \chi^2}} \\ &- \frac{2E_0(\mathbf{r})r^2 z \sin(\omega t - kz)}{\omega r_0^2 z_0^2(1 + z^2/z_0^2)\sqrt{1 + \chi^2}}, \end{aligned} \tag{5.11}$$

$$B_z(\mathbf{r}, t) = -\frac{2E_0(\mathbf{r})\chi x \cos(\omega t - kz)}{\omega r_0^2\sqrt{1 + \chi^2}} - \frac{2E_0(\mathbf{r})y \sin(\omega t - kz)}{\omega r_0^2\sqrt{1 + \chi^2}}. \tag{5.12}$$

To include the tunneling effect in ionization process, we allow the bound electron to tunnel through the potential barrier whenever it reaches the outer turning point, with a tunneling probability P^{tul} given by the WKB approximation [28, 30],

$$P^{\text{tul}} = \exp[-2\sqrt{2} \int_{\mathbf{r}_{\text{in}}}^{\mathbf{r}_{\text{out}}} \sqrt{V(\mathbf{r}_e) - V(\mathbf{r}_{\text{in}})} dl], \tag{5.13}$$

where $V(\mathbf{r}_{in}) = V(\mathbf{r}_{out})$, and the integration path is from $\mathbf{r}_{in}$ to $\mathbf{r}_{out}$ along the instantaneous direction of the electrical force. $V(\mathbf{r}_e) = -\frac{1}{|\mathbf{r}_e - \mathbf{r}_N|} + \mathbf{r}_e \cdot \mathbf{E}$.

In what follows, we exploit the above model to simulate the dynamics of accelerating neutral atoms of He and Mg. We will use hydrogen-like approximation and ionization potential parameters $I_p = 0.9$ and 0.28 correspond to He and Mg, respectively. The wavelength of laser is chosen as 800 nm and the waist of beam is 17.5 μm to match the experiment [24].

5.3 Distribution of Neutral Atoms and Their Accelerations

In the following simulations, we set the envelop of field ϵ_0 to be of trapezoid form, i.e., both the rise and the decay ramp last three cycles and the constant portion of the envelope lasts ten cycles. The moments when the constant portion begins and ends are denoted by t_B and t_E, respectively. The initial position of nucleus is located at the laser focal plane ($z = 0$) with $x = r_0/\sqrt{2}$ and $y = 0$. More than 500,000 samples were calculated for both He and Mg atoms in LP and CP fields.

5.3.1 Distribution of Neutral Atoms

After the laser pulse, we calculate the internal energy for each atom using the formula $E_i = m_r \dot{\mathbf{r}}_R^2/2 - 1/|\mathbf{r}_R|$, where the reduced mass $m_r = m_N/(m_N + 1)$ and the relative motion vector $\mathbf{r}_R = \mathbf{r}_e - \mathbf{r}_N$. We collect the neutral atoms whose energy is smaller than zero. The electrons of neutral atoms should stay in the elliptical Kepler orbit, whose semi-major axis b can be calculated according to $b = -1/2E_i$. We then make statistics on the neutral atoms according to their semi-major axes. The results are plotted as the black dotted lines in Fig. 5.1.

For He atom in LP laser field, almost 54 % of the total atoms survive after the pulse. Some electrons are excited into high Rydberg states with the elliptic orbits of the larger semi-major axes. Some stay in the ground state showing a peak in the distribution. The neutral atom distribution is closely related to the tunneling process. We find that almost 90 % of electrons in highly excited states experience tunneling process and the other 10 % of the electrons are excited into highly excited states without tunneling. According to our statistics, the ratio of the excited neutral atoms over the ionized atoms, i.e., He*/He+, is about 1 %. This ratio depends strongly on the laser intensity. We have lowered the intensity to 5×10^{14} w/cm² to match the parameters of the experiment [21], we find that the ratio value increases to 6 %, close to the value of 10 % in the experimental observation [21]. The deviation might be due to the trapezoid-form envelop of laser pulse used in our simulation, which differs from the Gaussian form envelop of the experiment and might produce relatively more ion yields. We adopt the trapezoid-form envelop here for convenience to calculate the

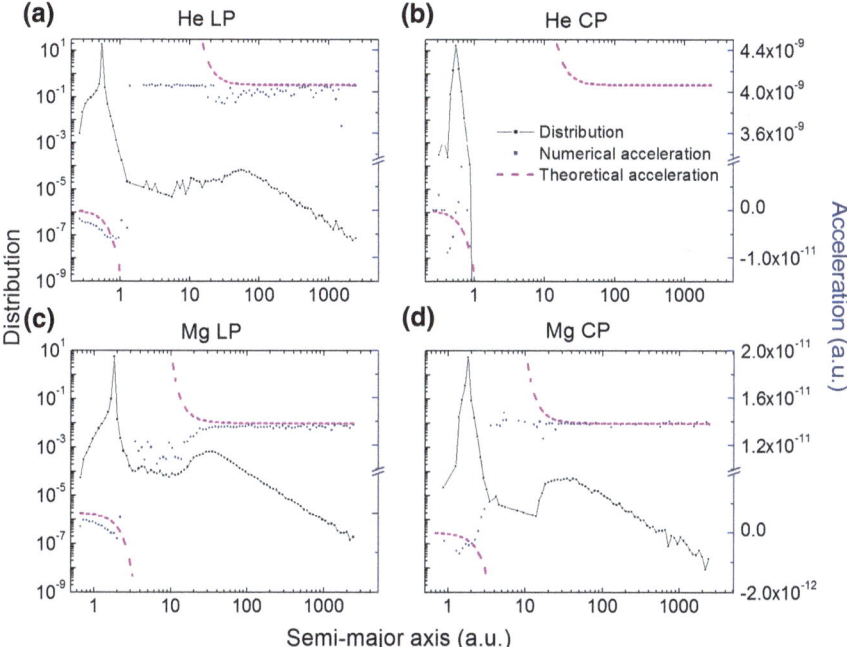

Fig. 5.1 The distribution of the neutral atom and the average acceleration after laser pulse. The *black dotted lines* denote the distribution of the electron orbit's semi-major axis of neutral atoms normalized to 1; *blue dots* denote the calculated average acceleration, while the discrete *purple curves* represent the results from the analytic formula (5.24). The laser peak intensity is 4.3×10^{15} w/cm^2 for He in (**a**) and (**b**), and 8.8×10^{13} w/cm^2 for Mg atoms in (**c**) and (**d**). Reprinted with permission from Ref. [27]. Copyright 2013, American Physical Society

atomic acceleration through the canonical velocity which is more properly defined for the laser field with a constant envelope, as will be shown in the following.

For He atom in CP laser field, however, we do not observe the electrons populated in highly excited states (refer to black dotted lines in Fig. 5.1). This result agrees with the experiment [21], which found that the excited-state atom yields decrease rapidly with the increase of the laser polarization ellipticity. The reason is that, in CP fields, the tunneled electron has a drift velocity that is proportional to the instantaneous electric field strength and its direction is perpendicular to the instantaneous polarized field direction. The drift velocity causes the electron to spiral out of its parent ion directly without return, which might greatly reduce the excited-state atom yields.

In contrast, the Mg atom distributions for both LP and CP cases exhibit analogous behavior as clearly shown by the black dotted lines in Fig. 5.1c, d, indicating that a significant number of the survival atoms populate in the highly excited states. This finding agrees with recent semiclassical results [31]. The reason is that the drift velocity of the tunneled electron can be compensated by its initial transverse velocity. The distribution of the initial transverse velocity takes the Gaussian form and its width is inversely proportional to the ionization potential. For Mg, the ionization potential is 0.28 a.u., which is much smaller than He of 0.9 a.u. Hence, the transverse velocity distribution is wide so that the initial transverse velocity can effectively cancel out the influence from the drift velocity. The tunneled electrons thus have a big chance to stay around the ion.

The peaks in all the above distributions indicate the high population of the survival atoms in the ground state. That is, the atoms of this part are neither ionized nor excited during the intense laser pulse. Moreover, in the following discussions we will show that they can be accelerated in the direction contrary to those highly excited atoms.

5.3.2 Average Atomic Acceleration

We calculate the acceleration $\mathbf{a}$ according to the canonical velocities of nucleus and electron, $\tilde{\mathbf{v}}_N(t)$ and $\tilde{\mathbf{v}}_e(t)$, respectively,

$$
\tilde{\mathbf{v}}_N(t) = \mathbf{v}_N(t) - \frac{E_0(\mathbf{r}_N(t))}{m_N \omega \sqrt{1 + \chi^2}} \sin(\omega t - k z_N(t)) \hat{e}_x
$$
$$
+ \frac{\chi E_0(\mathbf{r}_N(t))}{m_N \omega \sqrt{1 + \chi^2}} \cos(\omega t - k z_N(t)) \hat{e}_y, \tag{5.14}
$$
$$
\tilde{\mathbf{v}}_e(t) = \mathbf{v}_e(t) + \frac{E_0(\mathbf{r}_e(t))}{\omega \sqrt{1 + \chi^2}} \sin(\omega t - k z_e(t)) \hat{e}_x
$$
$$
- \frac{\chi E_0(\mathbf{r}_e(t))}{\omega \sqrt{1 + \chi^2}} \cos(\omega t - k z_e(t)) \hat{e}_y. \tag{5.15}
$$

Then the canonical velocity of mass center $\tilde{\mathbf{v}}_c(t) = (m_N \tilde{\mathbf{v}}_N(t) + \tilde{\mathbf{v}}_e(t))/(m_N + 1)$. The acceleration is then calculated as: (1) If the electron tunnels during the constant portion of the envelop, the average acceleration after tunneling is $\mathbf{a} = (\tilde{\mathbf{v}}_c(t_E) - \tilde{\mathbf{v}}_c(t_T))/(t_E - t_T)$; (2) Otherwise, we calculate the acceleration during the constant portion, i.e., $\mathbf{a} = (\tilde{\mathbf{v}}_c(t_E) - \tilde{\mathbf{v}}_c(t_B))/(t_E - t_B)$. Here, t_T is the tunneling moment, t_B and T_E are the moments when the constant portion begins and ends, respectively. We have made the average over the ensemble and the results are the blue dots in Fig. 5.1.

It is found that the average accelerations of the neutral atoms depend on electron states. Corresponding to the highly excited states, the atom can acquire large positive accelerations. For the He in LP, the value of the acceleration is around 4.1×10^{-9} a.u., i.e., 3.7×10^{13} g. Increasing the intensity to 10^{16} w/cm^2, the acceleration can be 10^{14} g which is comparable with experiment [24]. Corresponding to the ground state, the atom can acquire a small negative acceleration. To understand the average acceleration distribution, we make some deductions as follows.

We separate the atomic motions into the motion of mass center and the relative motion between nucleus and electron by introducing the following variables:

$$m = m_N + 1, \tag{5.16}$$

$$\mathbf{r}_c = \frac{m_N}{m}\mathbf{r}_N + \frac{m_e}{m}\mathbf{r}_e \approx \mathbf{r}_N, \tag{5.17}$$

$$\mathbf{r}_R = \mathbf{r}_e - \mathbf{r}_N \approx \mathbf{r}_e - \mathbf{r}_c, \tag{5.18}$$

For the first order approximation, we ignore the magnetic field force in the relative motion and obtain

$$\ddot{\mathbf{r}}_R \approx -\mathbf{E}(\mathbf{r}_c, t) - \frac{\mathbf{r}_R}{|r_R|^3} \tag{5.19}$$

Using $\mathbf{r}_N \approx \mathbf{r}_c$, the motion of the mass center is approximated by

$$m\ddot{\mathbf{r}}_c \approx -\mathbf{r}_R \cdot \nabla\mathbf{E}(\mathbf{r}_c, t) - \mathbf{v}_R \times \mathbf{B}(\mathbf{r}_c, t) \tag{5.20}$$

We use the harmonic oscillation motion to approximate the elliptic orbit motion governed by the Coulomb potential in Eq. (5.19). We have

$$\ddot{\mathbf{r}}_R \approx -\mathbf{E}(\mathbf{r}_c, t) - \omega_b^2 \mathbf{r}_R, \tag{5.21}$$

where the harmonic frequency (ω_b) is associated with the elliptic orbit semi-major axis (b) through $\omega_b = 1/b^{3/2}$.

Considering that the center motion is much slow compared to the harmonic motion, we can obtain an approximate solution for the above differential equation

$$\begin{aligned}
\mathbf{r}_R = &- \frac{E_0(\mathbf{r}_c)}{(\omega_b^2 - \omega^2)\sqrt{1 + \chi^2}} \cos(\omega t - kz_c)\hat{e}_x \\
&- \frac{\chi E_0(\mathbf{r}_c)}{(\omega_b^2 - \omega^2)\sqrt{1 + \chi^2}} \sin(\omega t - kz_c)\hat{e}_y.
\end{aligned} \tag{5.22}$$

Inserting the above solution into Eq. (5.21) and making average over the fast motion, after lengthy deduction, we obtain

$$m\ddot{\mathbf{r}}_c = \frac{1}{4}\frac{1}{\omega_b^2 - \omega^2}\nabla E_0^2(\mathbf{r}_c). \tag{5.23}$$

That is, the acceleration

$$\mathbf{a}_c = \frac{1}{4m}\frac{1}{\omega_b^2 - \omega^2}\nabla E_0^2(\mathbf{r}_c) \tag{5.24}$$

Note that, the above force can be viewed as a modified ponderomotive force. In the above deduction, the electric field term and the magnetic component are equally important and have been taken into account, or else the force may incorrectly appear to depend on the polarization of the field [32].

The above analytic expression is plotted as dashed purple curves in Fig. 5.1 and shows good agreement with our model calculations for the survival atoms in the regime of highly excited states and tightly bound states. In the resonant regime, i.e., $\omega_b \sim \omega$, the above formula diverges while the model calculations predict the finite average accelerations.

5.4 Maximum Velocity

The final velocity of atoms plays an important role in some applications such as chemical reaction, so that the maximum velocity is one of the major concerns in the acceleration process. To match experiment [24] and for convenience of deduction, we set the field envelop to be of Gaussian form, i.e., $\epsilon(t) = \epsilon_{peak} \exp(-t^2/\tau^2)$. According to the formula Eq. (5.24), the atoms located at the half beam size of $r = r_0/2$ in the laser field are expected to acquire the maximum acceleration. Integrating Eq. (5.24) over the full laser pulse, the maximum velocity $v_{max}(z)$ is

$$v_{max}(z) = \frac{\epsilon_{peak}^2}{4m\omega^2 w_0} \frac{\sqrt{\pi} e^{-0.5}}{(1 + (\frac{z}{z_0})^2)^{3/2}} \frac{\tau_{FWHM}}{\sqrt{\ln 2}}, \tag{5.25}$$

where FWHM means full-width at half-maximum and ϵ_{peak} is the peak field strength.

Equation (5.25) predicts that the atoms at the half-beam size on the focal plane of $z = 0$ will be accelerated to the largest velocity. Away from the focal plane along the direction of laser propagation, the atom will be less accelerated due to the decrease in laser intensity. Also, Eq. (5.25) implies linear relation between pulse duration τ_{FWHM} and maximum velocity.

References

1. Bulanov, S., Khoroshkov, V.: Plasma Phys. Rep. **28**, 453 (2002)
2. Fourkal, E., Li, J.S., Xiong, W., et al.: Phys. Med. Biol. **48**, 3977 (2003)
3. Fourkal, E., Shahine, B., Ding, M., Li, J.S., Tajima, T., Ma, C.-M.: Med. Phys. **29**, 2788 (2002)
4. Snavely, R., et al.: Phys. Rev. Lett. **85**, 2945 (2000)
5. Hegelich, M., et al.: Phys. Rev. Lett. **89**, 085002 (2002)
6. Hegelich, M., et al.: Phys. Plasmas **12**, 056314 (2005)
7. Leemans, W.P., Nagler, B., Gonsalves, A.J., Tó th, C., Nakamura, K., Geddes, C.G.R., Esarey, E., Schroeder, C.B., Hooker, S.M.: Nat. Phys. **2**, 696 (2006)
8. Esarey, E., Schroeder, C.B., Leemans, W.P.: Rev. Mod. Phys. **81**, 1229 (2009)
9. Mourou, G.A., Tajima, T., Bulanov, S.V.: Rev. Mod. Phys. **78**, 309 (2006)
10. Krausz, F., Ivanov, M.: Rev. Mod. Phys. **81**, 163 (2009)
11. McWilliams, C.M., Douglas, P., Barker, P.F.: Nat. Photon. **6**, 386 (2012)
12. Qiu, M., et al.: Science **311**, 144 (2006)
13. Gilijamse, J.J., Hoekstra, S., van de Meerakker, S.Y.T., Groenenboom, G.C., Meijer, G.: Science **313**, 1617 (2006)

14. Scharfenberg, L., van de Meerakker, S.Y.T., Meijer, G.: Phys. Chem. Chem. Phys. **13**, 8448 (2011)

15. Krems, R.V.: Phys. Chem. Chem. Phys. **10**, 4097 (2008)

16. Johnson, K.S., Thywissen, J.H., Dekker, N.H., Berggren, K.K., Chu, A.P., Younkin, R., Prentiss, M.: Science **280**, 1583 (1998)

17. Meschede, D.: J. Phys. Conf. Ser. **19**, 118 (2005)

18. Anderson, B.P., Kasevich, M.A.: Science **282**, 1686 (1998)

19. Tong, X.M., Zhao, Z.X., Lin, C.D.: Phys. Rev. A **66**, 033402 (2002)

20. Tong, X.M., Lin, C.D.: J. Phys. B: At. Mol. Opt. Phys. **38**, 2593 (2005)

21. Nubbemeyer, T., Gorling, K., Saenz, A., Eichmann, U., Sandner, W.: Phys. Rev. Lett. **101**, 233001 (2008)

22. Manschwetus, B., Nubbemeyer, T., Gorling, K., et al.: Phys. Rev. Lett. **102**, 113002 (2009)

23. Liu, H., et al.: Phys. Rev. Lett. **109**, 093001 (2012)

24. Eichmann, U., Nubbemeyer, T., Rottke, H., Sandner, W.: Nature **461**, 1261 (2009)

25. Fischer, M., Grossmann, F., Schmidt, R., Handt, J., Krause, S.M., Rost, J.M.: New J. Phys. **13**, 053019 (2011) Handt, J., Krause, S.M., Rost, J.M. et al.: arXiv:1103.1565 (2011)

26. Boot, H.A.H., Harvie, R.B.R.-S.: Nature **180**, 1187 (1957)

27. Xia, Q.Z., Fu, L.B., Liu, J.: Phys. Rev. A **87**, 033404 (2013)

28. Cohen, J.S.: Phys. Rev. A **64**, 043412 (2001)

29. Abrines, R., Percival, I.C.: Proc. Phys. Soc. London **88**, 861 (1966)

30. Ye, D.F., Liu, J.: Phys. Rev. A **81**, 043402 (2010)

31. Fu, L.B., Xin, G.G., Ye, D.F., Liu, J.: Phys. Rev. Lett. **108**, 103601 (2012)

32. Eberly, J.H., Javanainen, J., Rzazewski, K.: Phys. Rep. **204**, 331 (1991)

Chapter 6
Atomic Ionization in Relativistic Intense Laser Fields

Abstract In this chapter, we discuss atomic ionization in relativistic ultraintense laser. Our discussions are based on both analytical tunneling rate formula and numerical calculations using the classical trajectory Monte Carlo + tunneling (CTMC-T) approach. We address the quantum tunneling effect in the scheme of retrieving laser intensity via the ionization fraction of multiply charged ions. We also discuss the intensity dependence of photoelectron energy and momentum and the relevant carrier envelope phase (CEP) effect.

6.1 CTMC-T Calculation on Relativistic High-Z Atom Ionization

When the electron' quiver velocity (i.e., proportional to field strength over frequency) is comparable with light velocity, relativistic effect on atomic ionization is no longer negligible. Recent advances in optical technology have made this regime reachable: the peak intensity of ultrashort laser is up to 10^{22} W/cm^2 [1]. It opens a window to the physical phenomena occurring in relativistic and even ultra-relativistic regime [2].

6.1.1 CTMC-T Model

For the relativistic laser fields, the motion of electrons is governed by the following relativistic Newton–Lorentz equations:

$$\frac{d\mathbf{r}}{dt} = \frac{1}{\gamma}\mathbf{p}, \quad \frac{d\mathbf{p}}{dt} = -\left[\mathbf{E}(\mathbf{r},t) + \frac{Z\mathbf{r}}{r^3} + \frac{1}{\gamma}\mathbf{p}\times\mathbf{B}(\mathbf{r},t)\right], \quad (6.1)$$

with a relativistic factor $\gamma = \sqrt{1 + |\mathbf{p}|^2/c^2}$.

J. Liu, *Classical Trajectory Perspective of Atomic Ionization in Strong Laser Fields*,
SpringerBriefs in Physics, DOI: 10.1007/978-3-642-40549-5_6, © The Author(s) 2014

Numerical integration of the above differential equations can be performed over an ensemble of classical particles with initial energy equal to the ground state of high-Z ions $E_g = c^2\sqrt{1 - (Z/c)^2}$. The trajectory ensemble is assumed to be microcanonical and can be prepared for high-Z atoms as follows [3–5].

We first generate a trajectory ensemble distribution in the x–y plane, then extend to 3D space with the help of Euler angle rotation [6]. It can be readily proved that all possible initial states can be reached with choosing a parameter t_r randomly from the interval $\left[0, \Delta t = 2\pi/c^3 W^{3/2}\right]$ [7]. The parameter t_r relates to eccentric angle u of the Kepler orbit through the expression $u = c^3 W^{3/2} t_r + W^2 c^4 \bar{R} \Delta R \sin u$, which can be solved with iterative numerical technique. The eccentric angle determines the radius r and polar angle φ by the expressions $r = \bar{R} - \Delta R \cos u$ and $\varphi = 2\sqrt{\frac{s}{s-1}} \arctan\left[\sqrt{\frac{W}{s-1}} c^2 \left(\bar{R} + \Delta R\right) \tan \frac{u}{2}\right]$. Transforming back to the Cartesian coordinate, one obtains the initial positions $(x_0, y_0) = (r \cos u, r \sin u)$. The corresponding momenta (p_{x0}, p_{y0}) can be obtained by solving the equations for the ground state energy E_g and angular momentum L: $E_g = \sqrt{c^4 + (p_{x0}^2 + p_{y0}^2)c^2} - Z/\sqrt{x_0^2 + y_0^2}$, $L = x_0 p_{y0} - y_0 p_{x0}$. In the above expressions, we have used the abbreviations: $W = 1 - E_g^2/c^4$, $s = c^2 L^2/Z^2$, $\bar{R} = E_g/Wc^4$, and $\Delta R = \sqrt{1 - sW}/c^2 W$. Once the trajectory ensemble is prepared, each of them can be traced by numerically integrating Eq. (6.1) with the standard step adaptable Runge–Kutta algorithm. Moreover, quantum tunneling effect can be included by allowing the bound electron to tunnel through the potential barrier whenever it reaches the outer turning point, where $p_{i,z} = 0$ and $z_i \epsilon(t) < 0$. The tunneling probability P_i^{tul} is given by the WKB formula,

$$P_i^{\text{tul}} = \exp\left[-2\sqrt{2} \int_{z_i^{\text{in}}}^{z_i^{\text{out}}} \sqrt{V(z_i) - V(z_i^{\text{in}})} dz_i\right]. \tag{6.2}$$

Here, z_i^{in} and z_i^{out} are the two roots ($\left|z_i^{\text{out}}\right| > \left|z_i^{\text{in}}\right|$) of the equation for z_i, $V(z_i) = -Z/r_i + z_i \epsilon(t) = -Z/r_i^{\text{in}} + z_i^{\text{in}} \epsilon(t)$ [8].

6.1.2 Laser Intensity Dependency of High-Z Atom Ionization

We have applied CTMC-T approach to treat high-Z hydrogenlike atomic ionization in relativistic strong laser fields [3–5]. Figure 6.1 shows ionization fraction as a function of the peak laser intensity for hydrogenlike ions with different nuclear charge Z. The ionization curves exhibit a common feature: a flat profile near zero followed by a steep rise ending up with a plateau of complete ionization. Compared to the calculation from the pure classical approach without considering tunneling [9], all the curves shift toward low intensity regime by about one order of magnitude. It indicates that a laser field even far from the over-the-barrier regime can be strong enough to significantly ionize or even completely deplete the high-Z atoms through

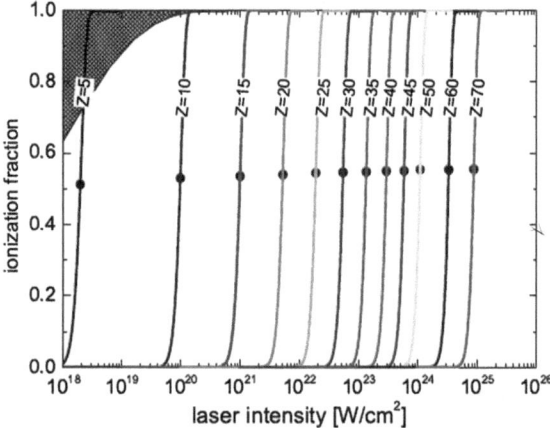

Fig. 6.1 The ionization fraction for several different hydrogen-like ions with nuclear charge Z as a function of the peak laser intensity. The ionization fraction is calculated at the end of a single-cycle planar pulse ($f(\eta) = 1$ if $0 < \eta < 2\pi/\omega$, otherwise $f(\eta) = 0$) of wavelength $\lambda = 1,054$ nm and $\phi_0 = 0$. The *hatched area* denotes the over-the-barrier ionization regime, other areas the tunneling regime. For a given ionic charge Z, the intensity threshold for over-the-barrier ionization can be calculated from Eq. (6.3). The connection of these critical points for different Z finally determines the boundary of the *hatched area*. The *full circles* are the most sensitive points, and are all located in the tunneling regime. Reprinted with permission from Ref. [3]. Copyright 2010, J. Phys. B

quantum tunneling. In Fig. 6.1, the hatched area denotes over-the-barrier ionization regime, the other part being tunneling regime. The boundary between the two regimes is given by

$$|\mathbf{E}_{bs}| = \frac{c^4}{4Z}\left(1 - \sqrt{1 - Z^2/c^2}\right)^2, \tag{6.3}$$

It is the relativistic generalization of the barrier suppression field strength at which the maximum of the effective Coulomb barrier equals the binding energy.

For a relativistic intense laser, the intensity measurement is a tough issue because traditional measurement [10] turns to be less viable in this regime [2, 11]. Instead of direct measurements, however, the sensitive dependence on laser intensity of the high-Z hydrogenlike ion yields can be exploited to characterize superintense laser field [9]. In Fig. 6.1, we can read out the most sensitive (MS) points on the ionization curves where the data change most steeply. We then collect data of ionization fraction and laser intensity corresponding to MS points for each curve and put them into Fig. 6.2. We also demonstrate the data calculated from the model without including tunneling for comparison. The curves of intensity versus ionic charge exhibit an analogous tendency, whereas the result from tunneling model is up to one-eighth lower than that from the model of no tunneling. On the other hand, the two curves for ionization fraction versus ionic charge manifest a complete difference: the calculation without tunneling shows a rapidly decrease as the ionic charge increases, whereas

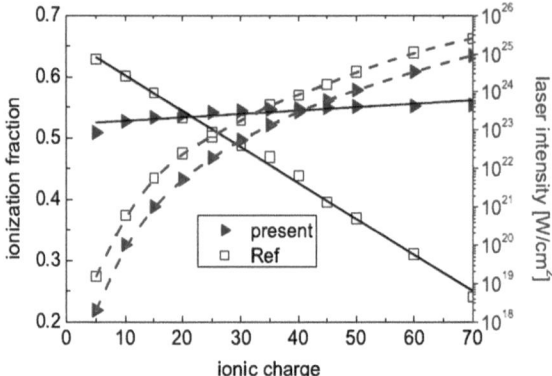

Fig. 6.2 The *solid line with full triangles* defines the most sensitively measured ionization fraction (*left axis*), whereas the *dashed line with full triangles* shows the corresponding laser intensity (*right axis*) as a function of the respective optimal Z. The *lines* are fits and the *triangles* indicate the deduced points from Fig. 6.1. All laser parameters are the same as those given in Fig. 6.1. The two *lines with open squares* are results without considering tunneling for comparison

the relativistic tunneling calculation predicts almost a straight line with slight ascent. The above comparison shows the significant effect of tunneling in relativistic regime, similar to that of nonrelativistic regime. With the help of the model calculation, one can retrieve the laser intensity in an indirect way. For instance, suppose the laser intensity waiting for measurement is roughly estimated to be $10^{23}\,\text{W/cm}^2$, then one must start from Z_{min} where the ionization fraction is near 1, and increase the ionic charge until Z_{max} where the ionization fraction is almost 0 (e.g. $Z_{min} = 20$, $Z_{max} = 40$). All the measured data should be put into the same figure, i.e., Fig. 6.2. The crossing of the measured curve (ionization fraction vs. ionic charge) and one of the reference curves (ionization fraction vs. ionic charge) gives the optimal ionic charge Z_{opt}, which finally determines the laser intensity according to the other reference curve (laser intensity vs. ionic charge).

6.2 Relativistic Semiclassical Tunneling Model

We now restrict our discussions to quasistatic tunneling regime where the relativistic generalized Keldysh parameter $\gamma = (\omega c/\,|\mathbf{E}_0|)\sqrt{1 - \left[\left(c^2 - Z^2/2\right)/c^2\right]^2}$ [12] is far below one. Here, $|\mathbf{E}_0|$ is the electric field strength, ω is the angular frequency, and $c = 137$ is the speed of light in atomic unit. The tunneling rate for the linearly polarized laser field in this regime has been obtained by Milosevic et al. through a semiclassical solution of the Dirac equation [13]

$$w_0(t_0) = \frac{c^2[|\mathbf{E}(\mathbf{r}_0, t_0)|/c^3]^{1-2\varepsilon}}{2\sqrt{3}\xi\,\Gamma(2\varepsilon+1)}\sqrt{\frac{3-\xi^2}{3+\xi^2}}\left[\frac{4\xi^3(3-\xi^2)^2}{\sqrt{3}(1+\xi^2)}\right]^{2\varepsilon}$$

$$\times \exp\left[6\mu\arcsin\frac{\xi}{\sqrt{3}} - \frac{2\sqrt{3}c^3\xi^3}{|\mathbf{E}(\mathbf{r}_0,t_0)|(1+\xi^2)}\right]. \tag{6.4}$$

Here, $\mu = Z/c$, $\varepsilon = \sqrt{1-\mu^2}$, $\xi = \sqrt{1-\varepsilon(\sqrt{\varepsilon^2+8}-\varepsilon)/2}$, and $\Gamma()$ is the standard Gamma function. $\mathbf{E}(\mathbf{r}_0, t_0)$ denotes the instantaneous laser electric field when the electron escapes out of the Coulomb barrier. In our following calculations [5], we adopt the rectangular Cartesian coordinate system and define axes Ox, Oy, and Oz as the magnetic, propagation, and polarization direction, respectively. The atom ensemble is placed on the origin so that $\mathbf{r}_0 \approx 0$. The laser field is a plane electromagnetic wave represented by the vector potential

$$\mathbf{A}(\mathbf{r}, t) = \frac{|\mathbf{E}_0|}{\omega}f(\eta)\sin(\omega\eta+\phi_0)\mathbf{e}_z, \tag{6.5}$$

with $f(\eta)$ the envelope function, $\eta = t - y/c$ the proper time, ϕ_0 the CEP, and $\mathbf{e}_z$ the unit vector in the z direction (the same for $\mathbf{e}_x$ and $\mathbf{e}_y$). The electromagnetic field can be obtained by means of

$$\mathbf{E} = -\frac{\partial\mathbf{A}}{\partial t} = -\frac{\partial A}{\partial t}\mathbf{e}_z, \tag{6.6}$$

$$\mathbf{B} = \nabla\times\mathbf{A} = \frac{\partial A}{\partial y}\mathbf{e}_x. \tag{6.7}$$

The ionization fraction is determined by integrating $w_0(t_0)$ over the time duration $[t_{\text{initial}}, t_{\text{final}}]$,

$$\chi = 1 - \exp\left[-\int_{t_{\text{initial}}}^{t_{\text{final}}} w_0(t_0)dt_0\right]. \tag{6.8}$$

To evaluate the photoelectron energy, we need to trace electron motion since they are released at the outer edge of the Coulomb barrier through quantum tunneling. In this stage, the classical mechanics is expected to work well. The motion of photoelectrons are then governed by the following relativistic Newton–Lorentz equations:

$$\frac{d\mathbf{r}}{dt} = \frac{1}{\gamma}\mathbf{p}, \quad \frac{d\mathbf{p}}{dt} = -\left[\mathbf{E}(\mathbf{r}, t) + \frac{1}{\gamma}\mathbf{p}\times\mathbf{B}(\mathbf{r}, t)\right], \tag{6.9}$$

with a relativistic factor $\gamma = \sqrt{1 + |\mathbf{p}|^2/c^2}$. It should be noted that in the relativistic regime the laser magnetic field significantly bends the classical trajectory and thus sharply suppresses the rescattering effect. As a result, we can safely neglect the nuclear Coulomb force in Eq. (6.9). We have also included the Coulomb force in our calculations, and confirmed that the rescattering effect can be neglected in the relativistic tunneling regime. With this approximation, the system is analytically solvable [14, 15]. It is easy to prove that there exist two constants of motion, i.e., $\mathbf{p}_\perp - \mathbf{A}$ and $c\gamma - p_\parallel$. Here, $\mathbf{p}_\perp$ and $p_\parallel$ are the components of the electron momentum perpendicular and parallel to the wave propagation direction, respectively. For electrons initially at rest, the second invariant in combination with the definition of γ [$\gamma = \sqrt{1 + |\mathbf{p}_\perp|^2/c^2 + p_\parallel^2/c^2}$], after some algebra, gives $|\mathbf{p}_\perp|^2 = 2cp_\parallel$. So, the electron kinetic energy measured by the detector outside the field can now be analytically determined as

$$E_{kin} = (\gamma - 1)c^2 = cp_\parallel = \frac{|\mathbf{p}_\perp|^2}{2} = \frac{|\mathbf{A}_0|^2}{2}. \tag{6.10}$$

Here, $\mathbf{A}_0$ is the instantaneous vector potential at the time when the electron is released through quantum tunneling.

In Fig. 6.3, we demonstrate a few-cycle pulse, time dependent ionization rate, and the final kinetic energy. The average energy over the pulse duration, i.e.,

$$\langle E \rangle = \frac{\int (dn/dt) E_{kin}(t) dt}{\chi}, \tag{6.11}$$

is plotted in Fig. 6.4.

The above results can be exploited to characterize the laser field with the following procedure. Before measurement, we should first estimate the applied laser intensity and select an appropriate species of ions according to Fig. 6.4. Then the ion ensembles are placed on the focus spot, and after irradiation the kinetic energies of the released

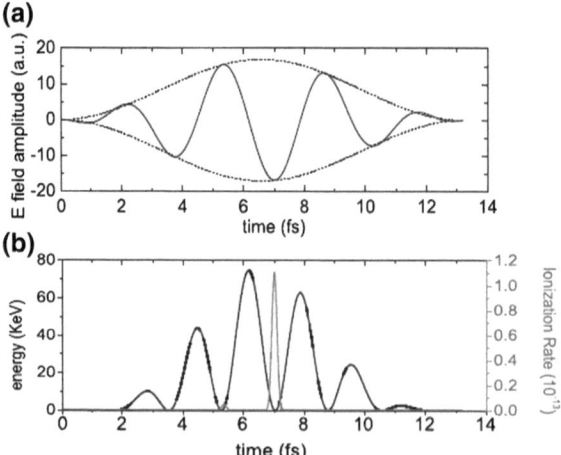

Fig. 6.3 **a** The electric field (*solid line*) and the envelope function (*dashed line*) of a 3.75-cycle laser pulse. **b** The ionization rate (*blue line*) and the final electron energy (*black line*) picked up from the laser field versus the ionization time

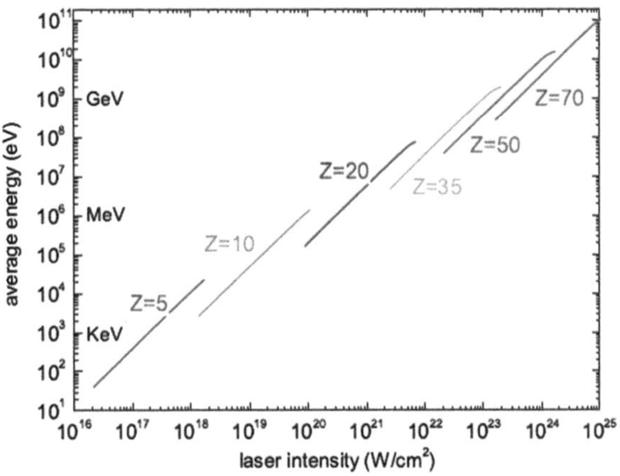

Fig. 6.4 The average
energy of photoelectrons
released from several different hydrogen-like ions
with nuclear charge Z as
a function of the peak
laser intensity. Other laser
parameters are: envelope function $f(\eta) = \sin^2(\pi\eta/2\tau_p)$,
full width at half maximum
FWHM) $\tau_p = 26.4$ fs, wavelength $\lambda = 1,054$ nm and
CEP $\phi_0 = 0$

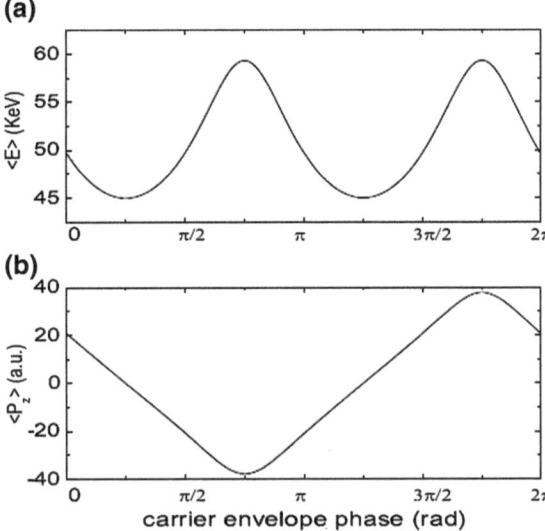

Fig. 6.5 a The average energy
of photoelectrons as function
of the CEP. We choose the
hydrogen-like ion as Ne^{9+}
($Z = 10$), the laser intensity is
10^{19} W/cm^2, the wavelength
is $\lambda = 1,054$ nm, and the
FWHM is $\tau_p = 6.6$ fs. **b**
Similar to **a**, but for the
averaged momentum along
the electric field direction

photoelectrons are recorded. The energies are averaged over all photoelectrons, and,
according to this averaged value, the laser intensity from Fig. 6.4. can be finally
read off. With the semiclassical model, we also calculate the CEP dependence of
average energy and momentum of photoelectrons. As shown in Fig. 6.5a, the average
energy of the photoelectrons presents sensitive dependence on the CEP varying from
45 to 60 KeV, and thus can be used to gauge the CEP. The same spirit has even
been adopted by a recent experiment that successfully demonstrates the single-shot
CEP measurement by comparing the number of photoelectrons emitted in opposing
directions parallel to the polarization of the laser at different energies [16].

References

1. Yanovsky, V., Chvykov, V., Kalinchenko, G., et al.: Opt. Express **16**, 2109 (2008)
2. Salamina, Y.I., et al.: Physics Reports **427**, 41 (2006)
3. Ye, D.F., Xin, G.G., Liu, J., He, X.T.: J. Phys. B **43**, 235601 (2010)
4. Ye, D.F.: Ph.D. Thesis, China Academy of Engeneering Physics (2011)
5. Xin, G.G.: Ph.D. Thesis, Beijing Institute of Technology (2012)
6. Cohen, J.S: Phys. Rev. A **26**, 3008 (1982)
7. Schmitz, H., Boucke, K., Kull, H.-J.: Phys. Rev. A **57**, 467 (1998)
8. Cohen, J.S.: Phys. Rev. A **64**, 043412 (2001)
9. Hetzheim, H.G., Keitel, C.H.: Phys. Rev. Lett. **102**, 083003 (2009)
10. Albert, O., et al.: Opt. Lett. **25**, 1125 (2000)
11. Mourou, G.A., Tajima, T., Bulanov, S.V.: Rev. Mod. Phys. **78**, 309 (2006)
12. Keldysh, V.: Sov. Phys. JETP **20**, 1307 (1965)
13. Milosevic, N., Krainov, V.P., Brabec, T.: Phys. Rev. Lett. **89**, 193001 (2002)
14. Sarachik, E.S., Schappert, G.T.: Phys. Rev. D **1**, 2738 (1970)
15. Krüger, J., Bovyn, M.: J. Phys. A **9**, 1841 (1976)
16. Wittmann, T., Horvath, B., Helml, W., Schätzel, M.G., Gu, X., Cavalieri, A.L., Paulus, G.G., Kienberger, R.: Nature Phys. **5**, 357 (2009)